Rapid Response Manufacturing

Contemporary methodologies, tools and technologies

Edited by

Jian (John) Dong

*Director of the Concurrent Design and Manufacturing Simulation
Laboratory (CDMS)
University of Connecticut
USA*

CHAPMAN & HALL

London · Weinheim · New York · Tokyo · Melbourne · Madras

Published by Chapman & Hall,
2–6 Boundary Row, London SE1 8HN, UK

Chapman & Hall, 2–6 Boundary Row, London SE1 8HN, UK

Chapman & Hall GmbH, Pappelallee 3, 69469 Weinheim, Germany

Chapman & Hall USA, 115 Fifth Avenue, New York NY 10003, USA

Chapman & Hall Japan, ITP-Japan, Kyowa Building, 3F, 2-2-1 Hirakawacho, Chiyoda-ku, Tokyo 102, Japan

Chapman & Hall Australia, 102 Dodds Street, South Melbourne, Victoria 3205, Australia

Chapman & Hall India, R. Seshadri, 32 Second Main Road, CIT East, Madras 600 035, India

First edition 1998

© 1998 Chapman & Hall

Typeset in 10/12 Palatino by Genesis Typesetting, Rochester, Kent

Printed in Great Britain by T. J. Press Ltd., Padstow, Cornwall

ISBN 0 412 78010 0

A catalogue record for this book is available from the British Library

Library of Congress Catalog Card Number: 97-75215

∞ Printed on permanent acid-free text paper, manufactured in accordance with ANSI/NISO Z39.48–1992 ANSI/NISO Z39.48–1984 (Permanence of Paper). [Paper = Magnum, 70gsm]*

To my parents: Qiai Dong and Daoqing Gao,
my wife: Jianmin Zhou
and my daughter: Catherine Z. Dong

Contents

Contributors

Mary Elizabeth A. Algeo
US Department of Commerce,
National Institute of Standards and Technology,
Manufacturing Systems Integration Division,
Gaithersburg,
Maryland 20899,
USA

M. D. Bauer
Systems Realization Laboratory,
The George W. Woodruff School of Mechanical Engineering,
Georgia Institute of Technology,
Atlanta,
GA 30332–0405,
USA

Marshall Burns
Ennex Fabrication Technologies,
10911 Weyburn Ave.,
Suite 332,
Los Angeles,
CA 90024,
USA

E. Caillaud
Ecole des Mines Albi-Carmaux,
Campus Jarlard,
Route de Teillet,
81013 Albi CT Cedex 09,
France

J. Dong
Space Systems Division,
MS AE44,
Boeing North American, Inc.,
12214 Lakewood Boulevard,
Downey,
CA 90242,
USA

John V. Draper
Robotics and Process Systems Division,
Oak Ridge National Laboratory,
POB 2008 MS 6304,
Oak Ridge,
TN 37831–6304,
USA

L. Felloni
Universita Di Ancona,
Department of Mechanics,
Via Brecce Bianche,
60131 Ancona,
Italy

J. E. Fowler
US Department of Commerce,
National Institute of Standards and Technology,
Manufacturing Systems Integration Division,
Gaithersburg,
Maryland 20899,
USA

A. Gatto
Universita Di Ancona,
Department of Mechanics,
Via Brecce Bianche,
60131 Ancona,
Italy

David C. Gossard
Computer Aided Design Laboratory
Department of Mechanical Engineering,
Massachusetts Institute of Technology,
Cambridge,
MA 02139,
USA

R. Ippolito
Politecnico di Torino,
Department of Production Systems and Economics
Corso Duca degli Abruzzi, 24, 10129
Torino,
Italy

L. Iuliano
Politecnico di Torino,
Department of Production Systems and Economics
Corso Duca degli Abruzzi, 24, 10129
Torino
Italy

Hugh Jack
Seymour & Esther Padnos School of Engineering,
Grand Valley State University,
L.V. Eberhard Center,
301 West Fulton, Suite 618,
Grand Rapids, Michigan 49504–6495,
USA

Kevin K. Jurrens
US Department of Commerce,
National Institute of Standards and Technology,
Manufacturing Systems Integration Division,
Gaithersburg,
Maryland 20899,
USA

Ali K. Kamrani
Industrial and Manufacturing System Engineering Department,
The University of Michigan–Dearborn,
Dearborn,
MI 48128–1491,
USA

Haeseong Jee
Department of Mechanical Engineering,
Hong-Ik University,
72-1 Sangsu, Mapo,
Seoul 121–791,
Korea

T. Manzur
Department of Mechanical Engineering,
University of Connecticut,
Storrs,
CT 06269–3139,
USA

S. H. Masood
Industrial Research Institute Swinburne (IRIS),
Swinburne University of Technology,
Hawthorn,
Melbourne,
Australia 3122

Daniel Noyes
Ecole Nationale d'Ingenieurs de Tarbes (ENIT),
Laboratoire Genie de Production,
Equipe Production Automatisée,
Avenue d'Azereix, BP 1629,
65016 Tarbes CEDEX,
France

David Rosen
Systems Realization Laboratory,
The George W. Woodruff School of Mechanical Engineering,
Georgia Institute of Technology,
Atlanta,
GA 30332–0405,
USA

C. Roychoudhuri
Department of Mechanical Engineering,
University of Connecticut,
Storrs,
CT 06269–3139,
USA

E. Sachs
Computer Aided Design Laboratory
Department of Mechanical Engineering,
Massachusetts Institute of Technology,
Cambridge,
MA 02139,
USA

P. R. Sferro
Industrial and Manufacturing System Engineering Department,
The University of Michigan–Dearborn,
Dearborn,
MI 48128–1491,
USA

Z. Siddique
Systems Realization Laboratory,
The George W. Woodruff School of Mechanical Engineering,
Georgia Institute of Technology,
Atlanta,
GA 30332–0405,
USA

Preface

In today's globally competitive manufacturing arena, the increasing customer demands for high-quality products delivered on time have forced companies to update their designs and make design changes more frequently than ever before. Rapid response manufacturing (RRM) has become a goal pursued by many companies to shorten time-to-market, improve quality-to-cost, and enhance product reliability.

There are strong needs for the research and development of technologies, methodologies and tools to carry out rapid response manufacturing (RRM). In the last few years, many new technologies have been developed for engineers to reduce the time required to design and manufacture products in response to rapidly fluctuating market demands. The chapters in this book address a variety of contemporary methodologies, technologies and tools for rapid response manufacturing. The contributions to this volume focus on two major RRM areas desktop manufacturing, computer and information technologies.

This volume consists of 12 chapters. In the second chapter Jee *et al.* present the research work on the development of the automated design of a three-dimensional printed mushroom surface texture which will be used in a rapid prototyping process, 3D printing. Rapid prototyping (or solid freeform fabrication) is one of the desktop manufacturing processes, which produce physical parts directly from computer models. In the third chapter Masood describes intelligent rapid prototyping by first highlighting the nature of the problems with current rapid prototyping systems and then discussing the configuration of the intelligent RP systems. Felloni *et al.* present an investigation of the surface features of rapid prototyping parts. Dong *et al.* discuss a new rapid prototyping technology by using fiber-coupled high-power laser diodes to directly sinter metal powders. Jurrens *et al.* describe the results from the rapid response manufacturing intramural project conducted at the National Institute of Standards and Technology (NIST), with specific emphasis on data standards for rapid response manufacturing. Kamrani and Sferro present the use of group technologies combined with knowledge-based systems for rapid response manufacturing. Bauer *et al.* describe the use of virtual prototyping technologies to aid the assessment of disassembly to put product life cycle concerns into the design stages. Draper discusses the human aspects of rapid response manufacturing. Caillaud and Noyes discuss the issues in fixture design from the point of view of CAD/CAM integration. Jack compares the use of solid freeform fabrication technologies and NC machining for rapid prototyping. Finally, the last chapter by Burns provides a view of future technologies in rapid response manufacturing.

As we look forward to the 21st century, we expect continuous growth and development of the RRM technologies. These technologies will gives companies more capabilities to quickly respond to the rapidly changing marketplace. The contributions to this book represent a foundation for the subsequent development of these technologies.

All authors and reviewers have contributed greatly their efforts to this book. I would like to take the opportunity to express my deepest thanks for their outstanding contributions and assistance in preparation of this book. My deep appreciation also goes to the book series editor, Dr Hamid Parsaei, for his endless support and brilliant encouragement. Finally, I am grateful to Mr Mark Hammond and Ms Deborah Millar of Chapman & Hall and Ms Joanne Jones, former employee of Chapman & Hall, for their patience and support throughout the project. It was a pleasure working with them.

Jian (John) Dong
December 1996

CHAPTER ONE

Introduction to rapid response manufacturing

J. Dong

This book discusses the methodologies, tools and technologies to implement rapid response manufacturing (RRM). RRM is a philosophy and concept which aims at reducing the time and cost of developing a high-quality product. In today's global competitive market place, manufacturers are continuously looking for technologies and methods to shorten the time and cost involved in product and process development and to ensure the quality of products. In the past ten years, numerous new technologies have been developed for rapid product design and realization, which has gained a lot of interest from government agencies, industries and academia.

The technologies for rapid response manufacturing involve research from a wide range of engineering and computer science areas. The current research and practice in rapid response manufacturing can be categorized into two major topics: desktop technologies for rapid response manufacturing, and computer and information technologies for rapid response manufacturing. This book documents the fundamental theories, current practice, and implementations of the two topics.

1.1 THE NEED FOR RAPID RESPONSE MANUFACTURING

To compete in the global marketplace, a company must keep updating their existing products and bring new products into the market. The process to develop an updated or a new product is sketched in Fig. 1.1.

With inputs from customers, design engineers and/or R&D engineers keep modifying their designs and/or creating new designs. The evaluations of functionality will be conducted afterwards. Often, a

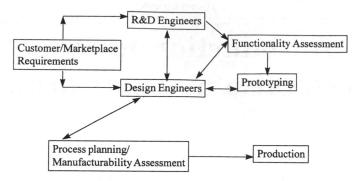

Fig. 1.1   Product development process sketch.

prototype will also be built. If the functionality is not satisfied, the design will be modified again.

At the same time process planners may conduct a manufacturability assessment, and create process plans for the updated or new product. When a part is designed, design engineers usually put design information into an engineering drawing or a CAD system, and then send the drawing or a CAD file to a manufacturing process planner. The process planner looks at the engineering drawing, and tries to interpret the design and capture as much design information as possible. Design data interpretation involves complicated human geometric reasoning processes. Based on the manufacturing capacities of a company, the process planner then uses his/her manufacturing knowledge to determine whether the part is manufacturable. If the part can be manufactured in the company at a reasonable cost, manufacturing process development will be performed, and process plans will be generated. If the part cannot be manufactured by the company or can only be manufactured at a high cost, the process planner will decide whether to send the part to an outside company to manufacture or to increase manufacturing capabilities by adding new machines, or ask designers to modify the design. After design modifications, the process planner will perform the same functions again.

However, in practice, this kind feedback rarely happens or may happen at very late stages. One reason is that design engineers have long been considered, or at least they consider themselves, superior to manufacturing engineers in manufacturing firms. The fact behind an engineering drawing is that manufacturing engineers (process planners) should use their knowledge to make the designed product producible. An engineering drawing (or a CAD drawing) is a good two-way communication tool among design engineers, but it is only a one-way communication tool between design engineers and manufacturing engineers. Because an engineering drawing does not provide quantitative manufacturing information, it is usually difficult or may take long time for process planners to convince designers to change their designs.

The iteration processes (design ↔ functionality assessment, design ↔ manufacturability assessment) are not only expensive but also time-consuming. It is therefore strongly recommended to consider both functionality and manufacturability at the early stage of design (Dong, 1996). There are currently two major approaches to integrated product and process development. One approach is to use management means to push design engineers with different disciplines to work more closely, for example, to form an integrated product and process team, or to put design engineers and manufacturing engineers in one room to physically make them communicate with each other easier.

Another approach is to rely on the technologies which make considera-tion of both functionality and manufacturability easier, and/or to eliminate some steps during the design and manufacturing iteration process. These technologies includes desktop manufacturing, and com-puter and information technologies.

1.2 THE APPROACHES AND TECHNOLOGIES FOR RAPID RESPONSE MANUFACTURING

1.2.1 Desktop manufacturing

Desktop manufacturing provides design engineers with easy and economic tools to produce a prototype or even a functional part in reduced time. The basic desktop manufacturing technologies include desktop numerical control (NC) machining, and solid freeform fabrica-tion (SFF). Both technologies aim at producing a physical part directly from a CAD model, and therefore production time and cost will be reduced. The essential components for a desktop NC machining system include a computer, CAD software, CAM software and a desktop machine, and the essential components for a SFF system include a computer, CAD software, slicing/path planning software and a SFF machine (Fig. 1.2).

Although the essential components for desktop NC machining and SFF are similar, the processes used are fundamentally different. Desktop NC machining mainly uses subtractive processes which include material removing. Because material removal includes significant cutting forces, the workpiece must be securely clamped, and usually a high level human involvement for planning and fixturing is required. A high-precision prototype or a functional part can easily be produced with desktop NC machining technology. In this book, Chapters 10 and 11 cover topics related to desktop NC manufacturing.

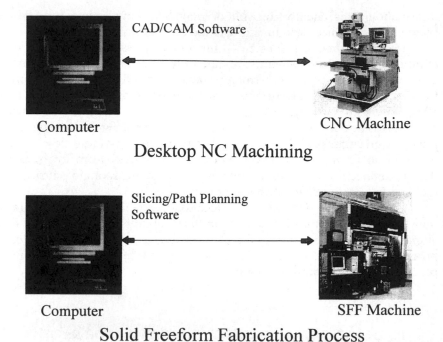

Fig. 1.2   Desktop manufacturing.

Solid freeform fabrication (SFF), an emerging desktop manufacturing process, has experienced dramatic growth in the last few years. Many SFF processes have been developed and new processes are emerging. Typical SFF processes include the following.

Laser-based SFF

This is one of the most popular SFF processes. Representative processes include the sterolithography process by 3D Systems Inc. (3D Systems 1988; McQuaid, 1994), and the selective laser sintering process by DTM Corp. A number of related processes, such as the diode laser (Dong, 1996), laser aided direct rapid prototyping (NSF, 1996), diode laser-based selective area sintering (DLSAS), and selected area laser deposition (SLAD), and selected area laser deposition vapor infiltration (SLADVI) processes (Dong *et al.*, 1996, Marcus *et al.*, 1996) are under development.

Shape deposition manufacturing (SDM)

The SDM method uses a layering technique to obtain near-net shape, intermingled with machining and other steps to refine the part shape and surface properties (NSF, 1996). Stratasys Inc. markets a machine using the technique. Further development of the SFF technique using the SDM method is conducted at a number of universities (NSF, 1996; Maxwell, 1996).

3D printing

This technology was developed at MIT. 3D printing functions by the deposition of powdered material in layers and the selective binding of the powder by a modulated 'ink-jet' printing of a binder material.

Laminate manufacturing

The representative process is the sheet-based process by Helisys Inc. Two fundamental steps are used in the process: cut a pattern and bond the pattern to the previous layer.

All SFF processes, however, are essentially additive processes (Burn, 1993), in which a physical part is built by adding materials. SFF processes are relatively young, and are expected to mature in their own way (Cohen *et al.*, 1995). About 75% of current SFF processes are primarily used for one-time prototype production, 20% is used for molding and investment casting, and less than 5% of SFF processes are used directly for the production of functional parts. The major advantages of using SFF technology are: (1) that the fabrication processes are directly driven by CAD models, and the entire product development time is therefore expected to be reduced; (2) the flexibility of the processes (theoretically, any given shape of part can be produced without tooling and fixturing); and (3) the SFF process promises to produce parts which cannot be produced with conventional methods, for example, non-homogeneous parts, and embedded electromechanical parts, etc. The disadvantages of using SFF are: that current SFF processes are expensive and the speed of production is low; and (2) that the structural integrity and surface finish of the part produced are low, compared with those produced by conventional manufacturing methods.

There are four common steps for current SFF processes. They are: geometric modeling and design evaluation, translating the geometric model into a STL file, (3) process planning, and (4) fabrication. Many research projects aimed at improving current SFF processes are currently conducted at both academia and industrial companies worldwide. This book covers some of them (Chapters 1–4).

1.2.2 Computer and information technologies for rapid response manufacturing

Rapid response manufacturing requires quick decision-making during product development stage. Correct and effective decision-making depends on information that accurately reflects the current status of a system and alternatives available throughout the design and manufacturing environment (Dong, 1995; Dunlap, 1987). The range of tools available to design engineers is steadily growing, which provides designers with the capabilities to compare different design alternatives and select the best design at lower cost and higher speed. With the development of computer network and communication technologies, people with different disciplines can work more closely and communicate more easily, which helps to provide accurate information quickly among engineers.

Although many design tools are available, most of these tools are used for functionality assessment and there is no available manufacturability assessment tool or even an algorithm (Dong, 1996). The issue has been widely recognized; many researchers have addressed the issue (Dong, 1996). In this book, Chapters 7, 8 and 10 cover the up-to-date research in the area.

Another major issue with these design tools is that the transformation of data among CAD, CAE and CAM tools is difficult because most of the tools were developed separately. A standard data format for information transfer among design engineers and manufacturing engineers is very important. Chapter 6 in this book has well addressed the issue. The human aspects for rapid response manufacturing are also addressed by Draper in Chapter 9.

The aim of the book is to represent the best accomplishments of researchers and system developers in these areas. This book will provide readers with an overview of methodologies, tools and technologies for rapid response manufacturing, as well as in-depth exposure to the current research issues and proposed solutions. Furthermore, the book covers various aspects of the applications of these methodologies, tools and technologies in real industry, and also identifies directions for future research and development. Readers will benefit from this text by gaining timely knowledge in the area of rapid response manufacturing.

REFERENCES

3D Systems Inc. (1988) Stereolithography interface specification, 3D system Inc., Valencia, CA, 1988.

Burns, M. (1993) *Automated fabrication: improving productivity in manufacturing*, Prentice-Hall, 1993.

Cohen, E., Drake, S., Gursoz, L., and Riesenfeld, R. (1995) Modeling issues in solid freeform fabrication. *NSF workshop on design methodology for solid freeform fabrication*, 1995.

Dong, J, Manzur, T., and Roychoudhuri, C. (1996) Fiber coupled high power laser diodes for solid freeform fabrication directly from metal powder. *Proceedings of the 1996 ASME-MED Winter Conference*, Atlanta.

Dong, J. (1996) The issues in computer modeling and interfaces to solid freeform Fabrication. *1996 ASME Winter Congress and Exposition*, Atlanta.

Dong, J., Parsaei, H., and Leep, H. (1996) Manufacturing process planning in concurrent design and manufacturing environment. *International Journal of Computers and Industrial Engineering*, **30**, (1), 83–93.

Dong, J. (1995) Feature-based interactive manufacturability assessment with process flow charts. *Proceedings of Concurrent Product and Process Engineering, 1995 ASME Winter Congress & Exposition*, San Francisco, CA, Vol. 1, pp. 329–337.

Dong, J. (1995) Organization structure, concurrent engineering and computerized enterprise integration. *International Journal of Concurrent Engineering, Research and Applications*, **3**, 167–176.

Dunlap, G. C. (1987) CIM – communication integrated manufacturing. *Proceedings of the 1987 Autofact Conference*, Detroit, MI.

Marcus, H., Harrison, S., and Cracker, J. (1996) Solid freeform fabrication: an overview. *Proceedings of the 1996 ASME-MED Winter Conference*, Atlanta.

Maxwell Project, 1996, http://www-personal.engin.umich.edu/~amarsan/Other/maxwell/maxwell.html.

McQuaid, J. (1994) Personal communication, SLC Format, 3D Systems Inc., Valencia, CA.

NSF (1996) NSF workshop on design methodologies for solid freedom fabrication, http://www.edrc.cmu.edu/proc/dmsff95.

CHAPTER TWO

Automated design of a three-dimensionally printed mushroom surface texture

Haeseong Jee, David C. Gossard and Emanuel Sachs

2.1 INTRODUCTION

2.1.1 Motivation for research

Three-dimensional printing (3DP) is a new solid freeform fabrication (SFF) process that creates parts directly from computer models [1]. 3DP functions by the deposition of powdered material in layers and the selective binding of the power by a modulated 'ink-jet' printing of a binder material. Following the sequential application of layers, the unbound powder is removed, resulting in a complex three-dimensional part, as illustrated in Fig. 2.1. 3DP is a highly flexible manufacturing process that allows for the fabrication of the component and assemblies of any shape in any material that can be obtained as a powder. The proper placement of binder droplets can be used to control the internal microstructure of the printed part and even create a component with local geometric and compositional control on a 100 µm scale.

Surface texture is a set of numerous tiny geometric features made in three dimensions on the surface of an object geometry for a physical application. It is significantly different from the textures in computer graphics that are widely used to add visual realism to a computer model since they were first suggested in 1974 [2]. Surface textures can be made for enhancement of heat and mass transfer and for tribological applications. For example, they can be made inside airfoils to increase the transfer of heat to coolant gasses. A surface texture in an orthopedic implant, on the other hand, helps the permanent fixation of the implant to the adjacent bone. Currently, the use of a porous surface texture on an orthopedic hip implant shown in Fig. 2.2 offers a valuable alternative to acrylic bone

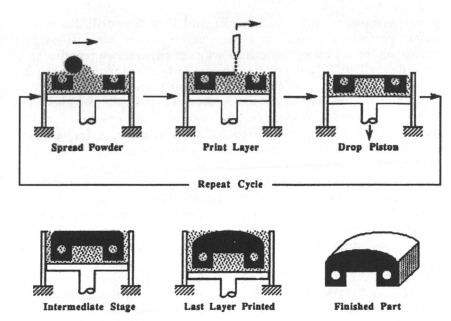

Fig. 2.1 The sequence of printing involved in 3DP.

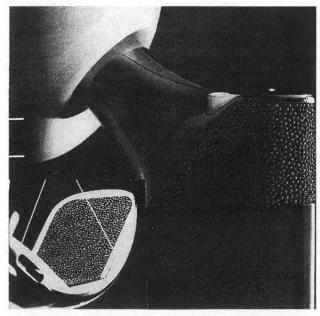

Fig. 2.2 A porous surface texture in an orthopedic hip implant.

cement as a means of fixation by promoting bone ingrowth. The specific surface porosity and pore size of the surface texture are thus essential conditions for providing bone cells with an environment conducive to osseointegration [3]. Depending on the desired feature geometry, size and material used, numerous surfacing techniques have been designed to attach or carve fine porous surface textures on orthopedic implants. Most techniques, however, affect the fatigue strength of the implant material [4], and are restricted to a specific surface texture with little room for variations of the surface parameters critical to the performance of the prosthesis. 3DP has the ability to fabricate a surface of controlled macro-textures with high geometric complexity, and a solid object that covered with a surface texture can be easily manufactured using a process illustrated in Fig. 2.3. This makes 3DP a promising alternative to fabricate surface textures for orthopedic implants [5].

Designing surface textures for 3DP, however, is difficult due to the non-traditional manufacturing method of the 3DP as well as the complex macro-structure of the texture geometry. As most current computer-aided design (CAD) systems are intended to fabricate a designed product geometry by traditional manufacturing technologies such as subtractive machining, forming or casting, they are not suitable to fully reflect the manufacturing limitations and flexibility of SFF technologies. Even most software efforts for SFF technologies have so far focused on ensuring compatibility between existing CAD tools and SFF processes, such as model repairing [6] or better model slicing [7,8], and will still be insufficient for exploiting the expanded design space offered by SFF technologies [9]. It is required to have an unprecedented design capability of handling a highly complex geometric model, which provides a manufacturable design without iterations by automatically taking the

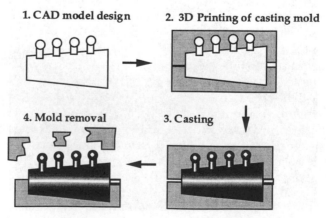

1. CAD model design **2. 3D Printing of casting mold**

4. Mold removal **3. Casting**

Fig. 2.3 The process of manufacturing surface texture using 3DP.

design rules into account. Specifically, two main problems arise, as follows. First, manufacturing has been an important source of constraints in product design. Manufacturability concerns must affect all stages of a design, and representing these concerns and understanding how they influence a design are of key importance [10]. Hence, the design of millimeter or sub-millimeter macro-texture features must fully reflect the manufacturing process rules of 3DP to be fabricated as accurately as designed. Second, surface textures are required to be regular on most complex object surfaces so that the regular pattern of a surface texture must be reconciled with the irregular surface curvature of most complex object surfaces. Creating undistorted regular textures has long been an important issue in generating realistic computer-generated images [11–13]. These, however, are only to add visual realism to CAD models, not to fabricate themselves. After all, we want to know how all the pieces of a surface texture geometry are supposed to fit together, whether the geometry of a designed texture feature still satisfies the design rules, and what a macro-texture feature that is being modeled is supposed to look like after manufacturing.

This chapter explores the application of an automated design paradigm in designing a surface texture to be fabricated by 3DP. It will demonstrate how this can be achieved by using a unified design constraint framework and an energy-based geometry. This method is tested through the implementation of a 'mushroom texture' field, an example of the surface texture.

2.1.2 Approach summary

The first effort to implement a CAD model and its real physical part with a surface texture has been made on quadric objects such as a sphere, conic cylinder or toroid, as shown in Fig. 2.4(a). Called a 'mushroom field', the texture consists of a regular array of elements distributed over a region of a component's external surface. The individual elements are 'mushrooms', consisting of a cylinder of a given radius whose axis is normal to the component surface topped by a sphere of a larger radius. The texture field is defined as a complex surface bound by an arbitrary NURBS curve defined interactively by the user in the (u, v) parametric domain. Once the texture field has been defined, the position of each single texture feature is carefully computed by considering the desired spacing and size of the texture feature set by the user. Those position points defined in the parametric domain are then mapped onto the real object domain to set the real position of each feature on the object surface. Thereafter, the template CAD model of one feature is copied into each previously calculated position, and the entire surface texture arrangement is constructed. After the CAD information of a part and texture has been well characterized, the model can then be directly translated into the

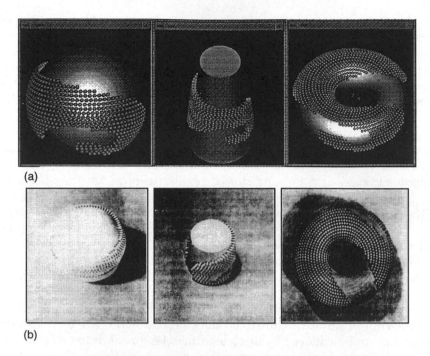

(a)

(b)

Fig. 2.4 The first implementation of designing and manufacturing surface textures by 3DP. (a) Three designed CAD models of the quadric object with a mushroom field. (b) Three 3D printed physical parts of the quadric object with a mushroom field.

manufacturing machine commands necessary to control the 3DP process. The physical fabricated parts of the CAD model using 3DP is shown in Fig. 2.4 (*b*). Even though this clearly shows the ability to design and manufacture a surface texture, however, it is only a demonstration or verification of the underlying concepts with a simple object geometry. Generalization of these concepts with necessary design constraints for a complex object geometry is thus still needed for further applications.

As the new approach for designing surface textures to be made by 3DP, an automated design paradigm using new design tools is proposed in this chapter. For surface texture applications, there are three important design requirements such as manufacturing process rules, regularity in texture mapping and functional requirements, as illustrated in Fig. 2.5. Among those three, functional requirements still remain yet undefined [14] and they are beyond the scope of the chapter. This chapter, however, proposes that the remaining two requirements for the surface texture design can automatically be satisfied by using the new design tools, which make satisfaction of functional requirements easier.

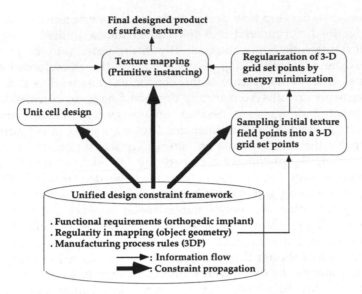

Fig. 2.5 A new design configuration of surface texture design.

The design process flow shown in Fig. 2.5 can briefly be summarized as the three main steps shown in Fig. 2.6. The first design step begins with creating a texture unit cell, defined as a geometric feature to be instantiated repetitively on an object surface to create a complete surface texture. For example, one mushroom feature shown in Fig. 2.6(*a*) is considered to be a texture unit cell. At the second step, a set of regularized lattice points can be acquired on a complex object surface (Fig. 2.6(*b*)), and, as the final step, the complete geometry of a surface texture is created by mapping the previous designed texture unit cell onto each lattice point on the surface (Fig. 2.6(*c*)).

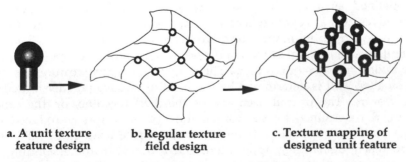

| a. A unit texture feature design | b. Regular texture field design | c. Texture mapping of designed unit feature |

Fig. 2.6 Three main steps of the surface texture design procedure.

In order to develop new design tools, first, manufacturing constraints are identified, formulated and introduced into a unified design constraint framework, being automatically incorporated into the process of the texture unit cell design (satisfaction of the manufacturing constraints). One unique feature of this constraint framework is that design requirements are all systematically organized so as to be considered in each relating design step. Second, an energy-minimizing parametric geometry is devised and implemented on a complex object surface. It will enable the user to create an optimal regular surface texture on the surface using the previously designed unit cell (satisfaction of mapping constraints) with texture mapping. Called the deformable lattice, this energy-minimizing geometry is built by the combination of a set of 3D lattice points plus an energy minimization analogy with an energy constraint. If the distance between any pair of lattice points is longer than the desired spacing, there is an elongation potential energy. On the contrary, if it is shorter than the desired spacing, there is a compression potential energy. In other words, the constraint functions play a role similar to that of physical energy functions during the constraint solving process, and the lattice geometry can automatically deform to minimize an energy functional subject to given geometric constraints on each lattice point. This is how all the geometric design parameters, including positions and orientations of the numerous surface texture features which satisfy the design requirements, can be found on most complex object surfaces.

2.2 MANUFACTURING CONSTRAINTS

The manufacturing constraint must specify the feasible dimensions and orientations of a product geometry in response to the manufacturing capability of a machine. As with many manufacturing processes, 3DP fields of application are also partly determined by the resolution, accuracy and minimum feature size it can make. In 3DP, features can be defined as a positive feature when created by an aggregate of primitives, or a negative feature when surrounded by a group of primitives. The minimum positive feature size is defined by the primitive size of single drops and lines of these single drops that form the building blocks of the parts. The minimum negative feature size can be smaller, as it is determined by the need to remove powder to define the feature. The overall accuracy of the 3DP machine, on the other hand, is determined by the combination of the errors introduced by each machine component. Manufacturing surface textures by 3DP is a particularly demanding application, and a thorough understanding of these manufacturing constraints is the crucial step for success.

2.2.1 Manufacturing process rules

In 3DP manufacturability is directly dominated by the manufacturing performance capability (MPC) of the machine, such as the minimum positive feature size, the minimum negative feature size and the minimum radius of curvature of a rounded feature. Though resolution is actually a limiting factor that needs to be quantified to determine the printability of very fine surface macro-textures, MPC is the most important manufacturability concern in design. In the meantime, MPC is characterized by the drop placement characteristics (DPC) of printing binder drops, such as the binder drop placement primitive size, printing accuracy and incremental

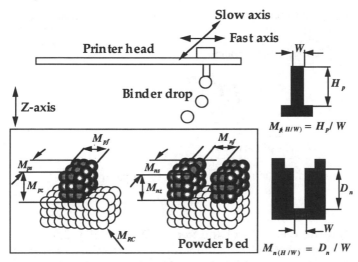

Fig. 2.7 The illustration of MPC in 3DP.

movements of binder drops along the three printing axes. A full set of MPC illustrated in Fig. 2.7 can be identified as follows:

M_{pf} minimum positive feature size along the fast axis
M_{nf} minimum negative feature size along the fast axis
M_{ps} minimum positive feature size along the slow axis
M_{ns} minimum negative feature size along the slow axis
M_{pz} maximum positive feature size along the Z axis
M_{nz} maximum negative feature size along the Z axis
$M_{p(H/W)}$ minimum ratio between width and height of a positive feature
$M_{n(H/W)}$ minimum ratio between width and height of a negative feature
M_{RC} minimum radius of curvature of a rounded feature

2.2.2 Formulation of manufacturing constraints

In designing surface texture, a vector set of the dimensional design parameters is represented as

$$\mathbf{D}_{DDP} = \{^{\exists} D_r^{A,B} | \, D_r^{A,B} \in \mathbf{R}^3, r = 1,2,3, \ldots, M\} \qquad (1)$$

where $\mathbf{R}^3$ is a subset of real values, M is the number of dimensional design parameters, and the notation of each parameter is illustrated in Fig. 2.8.

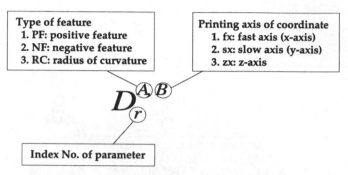

Fig. 2.8 The notations for a dimensional design parameter.

Based on the notations of the vector $\mathbf{D}_{DDP}$ in Fig. 2.8 and the MPC previously described, the manufacturing constraints can generally be described by defining $\mathbf{M}_{min}^{PF}$, $\mathbf{M}_{min}^{NF}$, $\mathbf{M}_{max}^{PF}$, $\mathbf{M}_{max}^{NF}$ and $\mathbf{M}^{RC}$ as the subsets of $\mathbf{D}_{DDP}$.

$$
\left.
\begin{aligned}
\mathbf{M}_{min}^{PF} &= \{^{\forall} D_r^{PF,fx}, \, ^{\forall} D_r^{PF,sx}, \, ^{\forall} D_r^{PF,zx} \in \mathbf{D}_{DDP} \,| \\
&\quad D_r^{PF,fx} \geq M_{pf}, D_r^{PF,sx} \geq M_{ps}, D_r^{PF,zx} \geq M_{pz}\} \\
\mathbf{M}_{min}^{NF} &= \{^{\forall} D_r^{NF,fx}, \, ^{\forall} D_r^{NF,sx}, \, ^{\forall} D_r^{NF,zx} \in \mathbf{D}_{DDP}| \\
&\quad D_r^{NF,fx} \geq M_{nf}, D_r^{NF,sx} \geq M_{ns}, D_r^{NF,zx} \geq M_{nz}\} \\
\mathbf{M}_{max}^{PF} &= \{^{\exists} D_i^{PF}, D_j^{PF}) \in \mathbf{D}_{DDP}|D_i^{PF}/D_j^{PF} \leq M_{p(H/W)}\} \\
\mathbf{M}_{max}^{NF} &= \{^{\exists} (D_i^{NF}, D_j^{NF}) \in \mathbf{D}_{DDP}|D_i^{NF}/D_j^{NF} \leq M_{n(H/W)}\} \\
\mathbf{M}^{RC} &= \{^{\exists} D_r^{RC} \in \mathbf{D}_{DDP}|D_r^{RC} \geq M_{RC}\}
\end{aligned}
\right\} \qquad (2)
$$

where $1 \leq i \leq R$, $1 \leq j \leq R$.

Another important manufacturing limit of 3DP is the staircase cusp drawn in Fig. 2.9. It is the typical machining mark of a layered manufacturing technology like SFF, which will limit the minimum

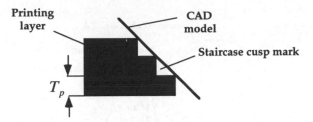

Fig. 2.9 An illustration of the staircase cusp mark in SFF techniques.

allowable feature size. This staircase mark can be formulated as $\mathbf{M}^{STA}$, another subset of $\mathbf{D}_{DDP}$, and is described as follows:

$$\mathbf{M}^{STA} = \{^{\forall}D_r^{PF}, {}^{\forall}D_r^{NF} \in \mathbf{D}_{DDP}|D_r^{PF} \geq R_{stm}^{P}, D_r^{NF} \geq R_{stm}^{n}\} \tag{3}$$

where R_{stm}^{p} is the minimum positive feature size constrained by the staircase mark, and R_{stm}^{n} is the minimum negative feature size constrained by the staircase mark.

2.2.3 Design rules by manufacturing constraints

In surface macro-texture design, simple design rules can be devised if only CSG solid primitives such as a box, cylinder and sphere are used to define a texture unit cell. Let us suppose that vectors of $k_1 M_{pf}$, $k_2 M_{ps}$ and $k_3 M_{pz}$, described in Section 2.2.1, where $k_1, k_2, k_3 \geq 1$ to ensure satisfaction of the manufacturing constraints, and draw these three along the three fabrication axes of 3DP, as shown in Fig. 2.10. An imaginary rectangular box can then be drawn based on these three orthogonal vectors. As each edge size of this box is exactly equal to that of $k_1 M_{pf}$, $k_2 M_{ps}$ and $k_3 M_{pz}$, three minimum positively fabricated feature sizes, this box represents a minimum 3D volume space that the 3DP machine can fabricate. A set of dimensional design parameters $L_{B,M}$, $W_{B,M}$ and $H_{B,M}$ for this box thus must be constrained as

$$\left. \begin{array}{l} L_{B,M} \geq k_1 M_{pf} \\ W_{B,M} \geq k_2 M_{ps} \\ H_{B,M} \geq k_3 M_{pz} \end{array} \right\} \tag{4}$$

As the orientation of a designed positive feature mapped onto a complex object surface generally does not coincide with the printing directions of the 3DP machine, the manufacturing constraints defined in (4) are not directly applicable to any designed feature in design process. An imaginary sphere is thus drawn by connecting imaginary paths of

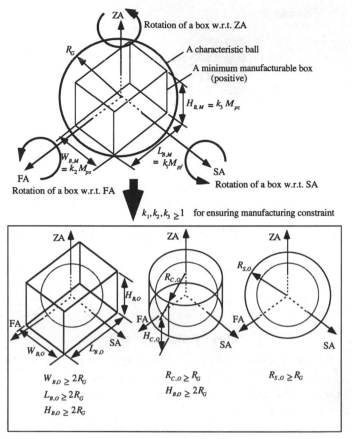

Fig. 2.10 Design rules on three CSG positive features by a characteristic ball.

eight vertex points of the box being rotated relative to its mass center. This ball then represents a minimum 3D volume space that can be designed to be fabricated by 3DP. According to the feature types (positive or negative), the radius of this ball can be easily determined as

$$R_G = R_{PB} = \frac{1}{2}\left[(k_1 M_{pf}^2) + (k_2 M_{ps})^2 + (k3 M_{pz})^2\right]^{\frac{1}{2}}$$

$$R_G = R_{NB} = \frac{1}{2}\left[(k_1 M_{nf}^2) + (k_2 M_{ns})^2 + (k_3 M_{nz})^2\right]^{\frac{1}{2}}$$

(5)

Again, k_1, k_2 and k_3 are proportional constants all greater than one to ensure the constraint satisfaction. A positive feature being mapped onto the complex object surface, for example, must be greater than or equal to $2R_{PB}$ to be fabricated by 3DP regardless of its orientation. As the result of these design rules, three CSG primitives on a complex object surface thus must be greater than the scale of the characteristic ball to satisfy the manufacturing constraints.

In addition the geometric feature on a complex object surface must also be constrained by a staircase mark constraint, as described in (3). After a simple geometry algebra, it turns out that

$$2R_{PB} = [(k_1 M_{pf})^2 + (K_2 M_{ps})^2 + (k_3 M_{pz})^2)]^{\frac{1}{2}}$$
$$\geq \max(M_{pf}, M_{ps}, M_{pz})$$
$$\geq \max(T_p, W_p)$$
$$\geq R_{stm}^P \tag{6}$$

and, similarly

$$2R_{NB} \geq R_{stm}^n \tag{7}$$

These, after all, show that the satisfaction of the staircase mark constraint is naturally guaranteed within prescribed manufacturing constraints, and only two characteristic ball diameters $2R_{PB}$ and $2R_{NB}$ must be considered as the key factors of the design rules to create CSG primitive features in texture unit cell design.

2.3 UNIFIED CONSTRAINT FRAMEWORK

2.3.1 A constraint configuration in unified framework

Figure 2.11 shows the configuration of the proposed unified constraint framework, where the three different design requirements seen in Fig. 2.5 can be systematically organized for the surface texture design. All the constraints, however, are not always activated in every design stage. The mapping and functional constraints, for example, are quite optional so that either or both of them may sometimes be deactivated. On the contrary, the manufacturing constraints are indispensable and must always remain active in this framework (black arrows for activated constraints and grey ones for deactivated constraints).

Another important characteristic in this framework is that the mapping constraints are given higher priority than functional ones in constraint satisfaction. It is not just because the latter remains yet undefined, but because the former must naturally take the precedence of the latter for a rational constraint propagation. Let us suppose two different constraint frameworks, illustrated in Fig. 2.12 for a comparison. In the case of Fig.

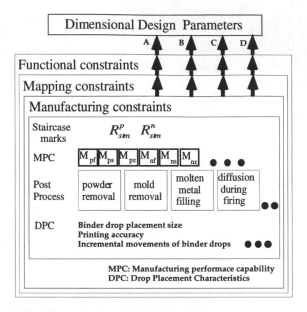

Fig. 2.11   A unified framework of design constraints.

2.12*a*, the design is supposed to satisfy functional constraints first and then mapping constraints later. In most cases of surface texture design, however, this will require unnecessary design iterations because achieving a functional goal must be the ultimate design intent of the surface texture design. This situation is more clearly visible in one statement, 'Surface texture needs an orthopedic implant stem to satisfy a functional requirement', which is obviously nonsense. This must be corrected to 'An orthopedic implant stem needs a surface texture to satisfy a functional requirement' to be feasible within the design intent for surface texture design. Hence, the case of Fig. 2.12(*b*) is a more natural way to construct a constraint framework that requires fewer iterations in the practical design procedure.

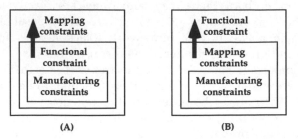

Fig. 2.12   Two design constraint frameworks.

2.3.2 Constraint satisfaction in unified framework

There are four different possible ways of manufacturing constraint satisfaction (*A, B, C, D*) into this framework. Each specific case of each of these is as follows:

- Case *A*: 1. Complex object surface
- Case *B*: 1. Complex object surface
- Case *C*: 1. A flat plane object surface
- Case *D*: 1. A flat plane object surface

2. Specific porosity and porous size
2. No functional requirement
2. Specific porosity and porous size
2. No functional requirement

For example, no mapping constraints exist (Cases *C* and *D*) when the object surface is simply a flat plane, whether or not it is slanted relative to the printing directions, and no functional constraints exist (Cases *B* and *D*) when the design is simply to test the manufacturing performance capability of the 3DP machine. Suppose that **J** is a feasible vector set of dimensional design parameters being defined within the unified constraint framework so that $\mathbf{J} \subset \mathbf{D}_{DDP}$. **J** can then be defined as one of the following four statements corresponding to the four different cases of constraint satisfaction.

$$
\left.
\begin{aligned}
&\text{Case } A\text{: } \mathbf{J} = \{^\exists D_r | D_r \in (\mathbf{M} \cap \mathbf{O} \cap \mathbf{F})\} \text{ to satisfy all design constraints} \\
&\text{Case } B\text{: } \mathbf{J} = \{^\exists D_r | D_r \in (\mathbf{M} \cap \mathbf{O})\} \text{ to satisfy the mapping constraints} \\
&\text{Case } C\text{: } \mathbf{J} = \{^\exists D_r | D_r \in (\mathbf{M} \cap \mathbf{F})\} \text{ to satisfy the functional constraints} \\
&\text{Case } D\text{: } \mathbf{J} = \{^\exists D_r | D_r \in \mathbf{M}\} \text{ to satisfy the manufacturing constraints only}
\end{aligned}
\right\} \quad (8)
$$

where **J** is a feasible subset of $\mathbf{D}_{DDP}$ within the constraint framework, **F** is a subset of $\mathbf{D}_{DDP}$ which satisfies the functional constraints, **O** is a subset of $\mathbf{D}_{DDP}$ which satisfies the mapping constraints, and **M** is a subset of $\mathbf{D}_{DDP}$ which satisfies manufacturing constraints.

For more specific examples, let us suppose that there is a set of four-dimensional design parameters D_1, D_2, D_3, D_4 defined within this constraint framework, and that each of these is identified as follows:

$$
\left.
\begin{aligned}
&D_1 \notin \mathbf{J} \\
&D_2 \in \mathbf{J} = \{^\exists D_r | D_r \in \mathbf{M}\} \text{ (Case } D) \\
&D_3 \in \mathbf{J} = \{^\exists D_r | D_r \in (\mathbf{M} \cap \mathbf{O})\} \text{ (Case } B) \\
&D_4 \in \mathbf{J} = \{^\exists D_r | D_r \in (\mathbf{M} \cap \mathbf{O} \cap \mathbf{F})\} \text{ (Case } A)
\end{aligned}
\right\} \quad (9)
$$

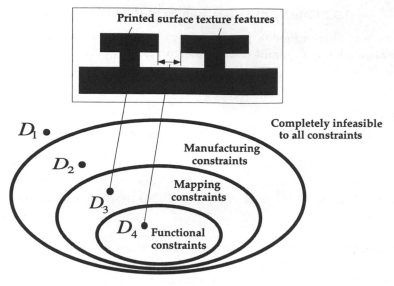

Fig. 2.13   An illustration of constraint satisfaction in the surface texture design.

A simple 2D surface texture model drawn in Fig. 2.13 then illustrates how the unified constraint framework will be used to decide the design constraint satisfaction. Suppose that this texture model is now defined by two-dimensional design parameters D_3 and D_4; one is for a positive feature and the other for a negative one. As illustrated in the figure, this texture feature then must be safely fabricated because D_3 and D_4 satisfy both the manufacturing and mapping constraints together. However, this surface texture will fail to work for the functional application because D_3 cannot satisfy the functional constraints.

2.4 DEFORMABLE LATTICE

2.4.1 Geometry

A deformable lattice is a set of regular lattice points in 3D space. Suppose that this set of points is defined by a set of N characteristic point vectors P_r, where $r = 1, \ldots, N$. If S_I denotes a subset of I point vectors in the same 3D space, the geometry is represented as follows:

$$S_N = \{P_1, P_2, \ldots, P_N\} \in \mathbf{R}^3 \tag{10}$$

The complete set of characteristic points in this lattice geometry, on the other hand, is also described by a geometry vector X, containing their Cartesian coordinates:

$$X = \{x_1, x_2, x_3, \ldots, x_{n-2}, x_{n-1}, x_n\}^T \tag{11}$$

where $n = 3N$; N is the number of characteristic points.

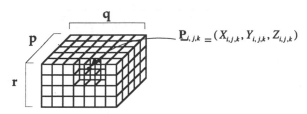

Fig. 2.14 A regular lattice that consists of a set of $N = p \times q \times r$ characteristic points.

If the lattice shown in Fig. 2.14 consists of N characteristic points where $N = p \times q \times r$, one complete set of characteristic points P_r can then be rewritten with a different notation $P_{i,j,k}$ according to their geometrical topology in this lattice. In that case

$$P_r = P_{i,j,k} \tag{12}$$

$$N = i + p(j - 1) + pq(k - 1) \tag{13}$$

where

$i = 1,2,3, \ldots, P, j = 1,2,3, \ldots, q, k = 1,2,3, \ldots, r, p$ is the number of lattice points in the i direction, q is the number of lattice points in the j direction, and r is the number of lattice points in the k direction.

Here, $P_{i,j,k}$ consists of three Cartesian variables, as follows:

$$P_{i,j,k} = X_{i,j,k}\hat{\imath} + Y_{i,j,k}\hat{\jmath} + Z_{i,j,k}\hat{k} \tag{14}$$

2.4.2 Total elastic energy

In this lattice geometry, a pair of two neighbouring points are considered to be connected with a linear elastic spring whose physical properties are characterized by a stiffness constant k and an initial length x_0 of the spring, as shown in Fig. 2.15. The elastic energy of a

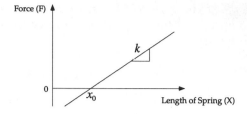

Fig. 2.15   The linear relation between the force and the length in a elastic spring.

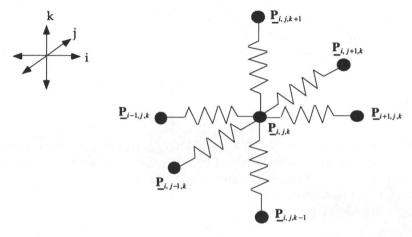

Fig. 2.16   Four neighbors of a selected node point vector $biP/bi_{i,j,k}$ in the lattice geometry.

lattice geometry can easily be determined by computing the spacing between a pair of neighboring lattice points. Suppose that $P_{i,j,k}$ shown in Fig. 2.16 is an arbitrary selected point vector inside a lattice. As all supposed lines that connect each pair of adjoined lattice points are now considered as linear elastic springs that have the same k and l_0, the elastic energy $U_{i,j,k}$ relative to the selected point $P_{i,j,k}$ can be determined as

$$
\begin{aligned}
U_{i,j,k} = \frac{1}{2}k\ [&(\|\overline{P_{i,j,k}P_{i-1,j,k}}\| - l_0)^2 + (\|\overline{P_{i,j,k}P_{i+1,j,k}}\| - l_0)^2 + \\
&(\|\overline{P_{i,j,k}P_{i,j-1,k}}\| - l_0)^2 + (\|\overline{P_{i,j,k}P_{i,j+1,k}}\| - l_0)^2 + \\
&(\|\overline{P_{i,j,k}P_{i,j,k-1}}\| - l_0)^2 + (\|\overline{P_{i,j,k}P_{i,j,k+1}}\| - l_0)^2]
\end{aligned}
\tag{15}
$$

and $\|\overline{P_{i,j,k}P_{i-1,j,k}}\|$ is the magnitude of the Cartesian vector $P_{i,j,k}P_{i-1,j,k}$ and represented as

$$\|\overline{P_{i,j,k}P_{i-1,j,k}}\| = [(X_{i,j,k} - X_{i-1,j,k})^2 + Y_{i,j,k} - Y_{i-1,j,k})^2 + Z_{i,j,k} - Z_{i-1,j,k})^2]^{\frac{1}{2}} \quad (16)$$

where $\|\overline{P_{i,j,k}P_{i-1,j,k}}\|$ is the magnitude of Cartesian vector $P_{i,j,k} - P_{i-1,j,k}$, k is the stiffness of linear elastic spring, l_o is the undeformed length of spring, $i = 1, \ldots, p, j = 1, \ldots, q, k = 1, \ldots, r$.

The total elastic energy of the lattice geometry U_E (X) then can be determined by summing the elastic energies relative to all lattice points. As each spring is shared by two neighbouring lattice points, its elastic energy will be summed twice in computing the total elastic energy. Therefore, the exact value of the total elastic energy $U_E(X)$ is determined by dividing the summed elastic energy

$$\sum_i \sum_j \sum_k U_{i,j,k}$$

by two

$$U_E (X) = \frac{1}{2} \sum_i \sum_j \sum_k U_{i,j,k} \quad (17)$$

2.4.3 Energy minimization

A set of N characteristic lattice point vectors which minimize (17) must be found. As each of these points now has three Cartesian coordinates, the minimization problem must be solved for $3N$ unknowns. However, (17) is a nonlinear equation and the number of design variables and constraints is quite large (the number of characteristic points in a lattice geometry used to design a custom surface texture is typically over one-thousand), analytical methods for solving this problem are either too complex or not applicable at all. Numerical methods are needed for the optimization technique to minimize the energy functional U_E (X).

All numerical methods to minimize an object function $U(X)$ are generally described by the following vector form of an iterative prescription [15]

$$\Delta X^{(t)} = X^{(t+1)} - X^{(t)} = \lambda_t \, d^{(t)} \quad (18)$$

where the subscript t denotes the design variable number, $X^{(0)}$ is any beginning design variable, and $\Delta X^{(t)}$ represents a small variation in current design variables for numerical optimization. A numerical method such as Newton's method would be extremely difficult to find all the $3N$ unknowns simultaneously, when N is a large number. For the energy minimization in this chapter, a gradient method has been used because it is simple and robust, though less efficient. The gradient vector at any point x_i, represents the direction of maximum increase in the function $U_E(X)$, and the rate of increase is the magnitude of the vector $\|U_E(X)\|$. The gradient vector of a scalar function $U_E(x_1, x_2, \ldots, x_n)$ can be defined as a column vector:

$$\nabla U_E = \left[\frac{\partial U_E}{\partial x_1} \quad \frac{\partial U_E}{\partial x_2} \cdots \frac{\partial U_E}{\partial x_p} \right] \equiv g \tag{19}$$

and in (18)

$$d = -g \tag{20}$$

2.4.4 Projection of lattice points

The set of lattice point vectors is required to be constrained onto the surface geometry of object models. In other words, all lattice points must always be positioned on the object surface during the process of energy minimization. The surface geometry of most object models, however, is a NURBS surface given explicitly by its parametric equation $f(u,v)$ and will not be generalized as a function of the design variable vector $X = (x_1, x_2, \ldots, x_n)$. Therefore, a typical Lagrange multiplier [16] to solve this geometrically constrained minimization problem is not at all applicable.

In this chapter, the constrained minimization process consists of a technical combination of two numerical methods: an energy minimization plus the projections of lattice points back to the surface. As illustrated in Fig. 2.17, the first stage is the simple unconstrained energy minimiza-

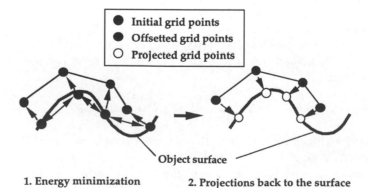

Fig. 2.17 Illustration of the two-stage constrained energy minimization process of a deformable lattice on an object geometry.

tion by the gradient method. As unconstrained, however, the lattice points will be offset from the surface of the object model. In the second stage, therefore, these offset lattice points must be projected back to the object surface. In short, the energy minimization problem will first be solved with a step of unconstrained minimization, followed by the next step of point projection onto the object surface. The lattice points thus will always be kept on the surface of an object geometry during the process of energy minimization. This two-stage numerical process of energy minimization will be continued over and over again until the termination conditions are met, which means a fully minimized object energy functional.

2.5 AUTOMATED DESIGN OF MUSHROOM SURFACE TEXTURE

An implementation has been made using the automated CAD system, which was developed on a SilconGraphics Indigo 2 engineering work-station at MIT CADLAB. As mentioned before, the first stage of design begins with the creation of a mushroom texture unit cell based on design rules. Then the surface of an object geometry is regularly sampled into a set of lattice points, the post-process for the regularization process, and each pair of neighboring lattice points is constrained by the energy constraint. The elastic potential energies of all the pairs of lattice points are then summed to a scalar energy functional, to be minimized later by a numerical method. As a result, a set of regularized lattice points can be acquired on a complex object surface, and then the complete geometry of a surface texture is finally created by mapping the previously designed mushroom cell onto each lattice point on the surface.

2.5.1 Design of a mushroom texture unit cell

A mushroom texture unit cell has been designed as the first step of the automated design. It consists of two simple CSG primitives, a sphere and

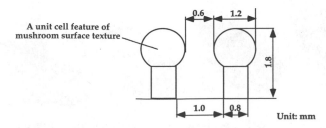

Fig. 2.18 Dimensional design constraints on a mushroom surface texture.

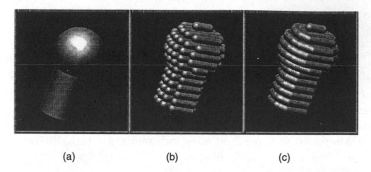

(a)   (b)   (c)

Fig. 2.19   An example of visual 3DP simulation of the mushroom texture feature.

a cylinder, whose dimensional parameters are limited by prescribed design rules as specified in Fig. 2.18. To verify the aesthetic rule, in the mean time, the CAD model of a mushroom texture feature (Fig. 2.19(*a*)) has been inspected through visually simulating the fabrication based on two different 3DP techniques currently being used (Fig. 2.19(*b*) and (*c*)). The result of visual simulation can also be compared later with the real physical part (Fig. 2.20) fabricated by 3DP.

Fig. 2.20   The real physical shape of a printed mushroom surface texture made by 3DP.

2.5.2 Sampling initial lattice points for a texture field

A NURBS surface, chosen as a surface geometry of an object model, will first be regularly sampled into a set of grid lattice points ($19 \times 19 = 361$) along the isoparametrics in the parametric domain. This grid lattice will be the initial configuration of the lattice points for a further numerical process of energy minimization.

2.5.3 Regularization of lattice set points by energy minimization

As the orientation of each mushroom is exactly the normal to the component freeform surface, however, the regularity in the top will not be the same as that in the bottom. To overcome this problem, a two-layer grid lattice points set is generated on the same NURBS surface. The difference in regularity between these two is illustrated in Fig. 2.21. After

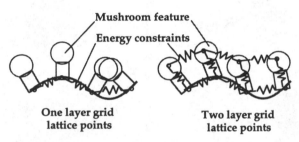

One layer grid
lattice points

Two layer grid
lattice points

Fig. 2.21 Illustration of the regularity difference between two mushroom surface textures.

several numerical steps of the energy minimization, the regular lattice points set in either case can be found (Fig. 2.22; middle row) on the object surface after the two-stage energy minimization described in Section 2.4.4.

2.5.4 Texture mapping

As the last step, each of the lattice set points on the object surface can be occupied, one-by-one, through primitive instancing of the previously defined mushroom texture unit cell until a complete surface texture geometry has been made (Fig. 2.22; bottom row). Each cell primitive on the surface has its own orientation, depending on the surface geometry. As expected before, the mushroom surface texture using a two-layer grid lattice (Fig. 2.22; right column) shows a more regular pattern on the top and the bottom.

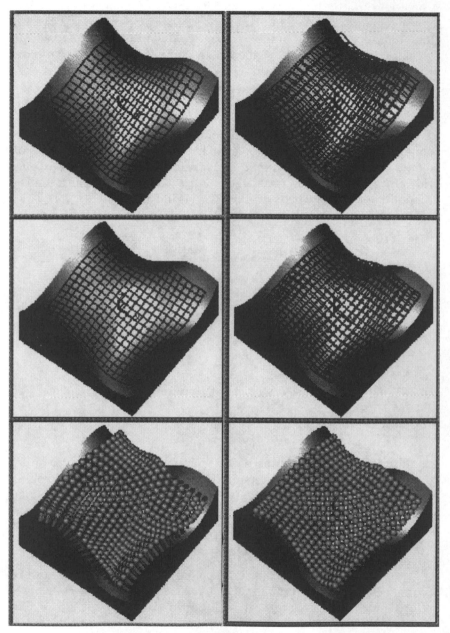

Fig. 2.22   Two mushroom surface textures made on the object NURBS surface using a one-layer grid lattice (left) and a two-layer grid lattice.

2.6 CONCLUSION

The approach in this chapter is to develop a new automated computer-aided design tool which enables the designer to create an unusual solid geometry, surface texture, which will be manufactured accurately by three-dimensional printing, one of the solid freeform technologies. An energy-based geometry, the deformable lattice, has been devised by the combination of a parametrically described lattice geometry plus an energy minimization strategy that acts to optimize the surface texture geometry in response to the internal and external geometric design constraints. Also, a unified design constraint framework with design rules will keep the designed texture geometry from violating the manufacturing constraints. This approach is tested by designing a three-dimensionally printed mushroom surface texture and will be able to make contributions towards future surface texture designs in the following ways.

2.6.1 Improvement in user interactivity

Being capable of modifying itself in response to given geometric constraints, it greatly reduces the amount of management input required to define and modify the shape of the product geometry.

2.6.2 Enlargement of designability

Although conventional methods return no solution to a geometrically overconstrained problem, this method provides the designer with an optimized design solution, a compromise that is sometimes reasonably acceptable.

2.6.3 Conservation of manufacturability

This can always keep the design solution within the feasible range of product manufacturability by incorporating manufacturing constraints into the design process.

As a future direction, carefully defined topologies between design parameters and well-chosen numerical algorithms will help this approach attain useful design solutions to manufacture surface textures quickly and easily.

ACKNOWLEDGMENTS

This work is supported by the NSF Strategic Manufacturing Initiative Grant DDM 9215728. Support by the 3D Printing Industrial Consortium, ARAP, Therics, Inc., and the MIT Leaders for Manufacturing Program is gratefully acknowledged.

REFERENCES

[1] Sachs, E. *et al.* (1992) Three dimensional printing: rapid tooling and prototypes directly from a CAD model. *Journal of Engineering for Industry,* November.

[2] Catmull, E. E. (1974) *A subdivision algorithm for computer display of curved surfaces.* PhD thesis, Department of Computer Science, University of Utah.

[3] Bobyn, J. D. *et al.* (1980) The optimum pore size for the fixation of porous-surfaced metal implants by the ingrowth of bone. *Clinical Orthopaedics,* **150,** 263–270.

[4] Callaghan, J. J. (1993) Current concepts review: the clinical results and basic science of total hip arthroplasty with porous-coated prostheses. *Journal of Bone and Joint Surgery,* **75-A,** (2).

[5] Jee, H. (1996) Computer-aided design of surface macro-texture for three dimensional printing. PhD thesis, Mechanical Engineering Department, MIT, 1996.

[6] Bøhn, J. H., and Wozny, M. J. (1992) Automatic CAD-model repair. *4th IFIP 5.2 Workshop on Geometric Modeling in Computer-Aided Design,* RPI, New York.

[7] Suh, Y. S., and Wozny, M. J. (1994) Adaptive slicing for solid freeform fabrication process. *SFF Symposium '94,* University of Texas, Austin.

[8] Dolenc, A., and Mäkelä, I. (1994) Slicing procedures for layered manufacturing techniques. *Computer-Aided Design,* **26,** (2).

[9] Chandru, V., Manohar, S., and Prakash, C. E. (1995) Voxel-based modeling for layered manufacturing. *IEEE Computer Graphics and Applications.*

[10] Arbab, F. (1990) Features and geometric reasoning. *Intelligent CAD, II, IFIP,* Elsevier Science Publishers, North-Holland.

[11] Bier, E. A., and Sloan, Jr., K. R. (1986) Two-part texture mappings. *IEEE Computer Graphics and Applications.*

[12] Ma, S. D., and Lin, H. (1988) Optimal texture mapping. *EUROGRAPHICS '88,* North-Holland.

[13] Bennis, C., Vézien, J.-M., and Iglésias, G. (1991) Piecewise surface flattening for non-distorted texture mapping. *Computer Graphics, SIGGRAPH,* **25,** (4).

[14] Curodeau, A. (1995) *Three dimensional printing of ceramic molds with accurate surface macro-textures for investment casting of orthopaedic implants.* PhD thesis, Mechanical Engineering Department, MIT, 1995.

[15] Arora, J. S. (1989) *Introduction to optimum design,* McGraw-Hill Series on Mechanical Engineering.

[16] Bertsekas, D. P. (1976) Multiplier methods: a survey. *Automatica,* **12.**

CHAPTER THREE

Intelligent rapid prototyping

S. H. Masood

3.1 INTRODUCTION

Rapid prototyping is a new development in manufacturing technology that involves a group of manufacturing techniques that are based on layer-by-layer material deposition rather than on material removal or deformation. The technology enables rapid production of complex 3D objects directly from computer-aided design model without involving any conventional tooling or numerical control part programming. Systems are in use by many companies to make models and prototypes for focus group evaluation, testings and downstream moulding and casting processes. Rapid prototyping greatly reduces the time and cost necessary to bring new products into the market. Its ability to make many successive models of a product easily will greatly enhance the capability of the designer to optimize the functionality, appeal and quality of their products.

Over the past few years, rapid prototyping techniques have become commercially available and are now having a significant impact on the overall design-production cycle in industry. Dominated by stereolithography, around 1000 rapid prototyping machines came into use by mid-1995, and the number is increasing steadily [1]. Service bureaus that offer users the opportunity to try out new technologies and test their products are beginning to appear in large numbers, especially around industrial manufacturing regions, where some 90% of all prototyping currently takes place.

Presently most of the basic research work in rapid prototyping is directed towards the development of new materials or techniques for material deposition [2]. New application-based research is needed in several areas to make this technology more efficient, cost-effective and versatile, leading to further improvements in the product development cycle. There are a number of issues that now confront the industry in the development, implementation and application of rapid prototyping (RP) technology. Some of the main issues are:

- integrating RP systems with modern design tools such as feature based modeling
- increasing efficiency and flexibility of RP processes with intelligent process planning
- improving the quality of parts and speeding the process with intelligent fixture design
- achieving a balance in conflicting design requirements in product development.

This chapter attempts to address these issues by presenting a framework for an intelligent rapid prototyping system based on the best attributes of available modern core technologies of knowledge-based systems, object-oriented programming, feature-based modeling and distributed black-board technology. The main objective is to resolve the problems associated with the design, costing, process planning, fixturing and inspection of rapid prototyping parts, and thereby significantly improve the product development cycle in terms of speed, economy, accuracy and versatility of the process.

The chapter focuses on a distributed blackboard approach, which provides a cooperative environment to facilitate the exchange of information and data between each of the associated knowledge-based systems. The chapter highlights the nature of the problems with the current RP system, and then discusses the configuration of the intelligent RP system and the associated knowledge-based systems for design, process planning, fixturing and value engineering. The present work is not directed to any of the specific rapid prototyping technologies available, and therefore the approach is applicable in general to any rapid prototyping system

3.2 ISSUES WITH CURRENT RAPID PROTOTYPING PROCESSES

Researchers at universities, research centres and corporations around the world have developed or are in the process of developing rapid prototyping technologies. At least six of these technologies are commercially available. Among these, stereolithography, developed by 3D Systems Inc., was the first rapid prototyping system introduced in 1987. All the commercial systems are based on the concept of building a part by adding thin slices layer-by-layer using the sliced model from a computer-aided design system. The six major rapid prototyping technologies (with the names of their manufacturers) are:

- stereolithography (SLA)   3D Systems Inc.
- fused deposition modeling (FDM)   Stratasys Inc.
- laminated object manufacturing (LOM)   Helisys Inc.
- selective laser sintering (SLS)   DTM Inc.
- solid ground curing (SGC)   Cubital Inc.
- 3D plotting   Sanders Prototype Inc.

Each of the above technologies involves different methods of generating slices, uses different materials, and offers various advantages, limitations and applications. Several studies have been made of comparing and evaluating different rapid prototyping processes [3,4].

Common to all rapid prototyping processes is the basic requirement of CAD modeling and preparing the model in a form acceptable to each system. A CAD model of the object to be produced is first created as a solid model or a closed volume surface model. This ensures that all horizontal cross-sections are closed curves which create a solid object. The solid model or surface model is then converted into a format called an STL file by a software supplied by the CAD vendor. The STL file approximates the surfaces of the object by the tessellation of small triangles. Another software analyses the STL file and then slices the model into thin cross-sections. It is these cross-sections which are created as layers and build up the part by the specific rapid prototyping technology. As the part is created by combining layers, there are various choices of position and orientation in which the same part can be built. An appropriate part positioning and orientation are necessary in all rapid prototyping processes for a number of reasons. These include the accuracy, appearance, surface finish, functionality, cost and time of production. There are no specific rules for this and it depends a great deal on the experience and judgement of the designer.

In many RP systems, the part is built on fixtures which are created either at the time of CAD modeling or are generated by the RP software during preprocessing. These fixtures are physically removed from the part after the part building is finished. The creation, selection and removal of fixtures also affect the quality and cost of the parts built. In the SLA process, the building of fixtures is essential to support the part. Fixtures may also be required in FDM and other processes. Again, the selection of shape, size and positioning of these fixtures depends greatly on the judgement of the part designer.

In addition, there are several other difficulties and productivity problems that are confronted by the experienced users of the RP processes, which necessitate several knowledge-intensive activities to be performed during and after the part design process. The key activities needed prior to the RP building process include the following.

Part design involves the determination of the geometrical attributes and qualitative properties such as the required aesthetic, functionality, surface finish, allowable costs and delivery lead-time.

Process planning encompasses the configuration of a specific orientation to achieve desired surface finish and reduce the set-up and secondary operation costs and laser costs (as in SLA and SLS).

Fixturing design involves the development of the necessary supports to minimize the amount of distortion and potential defects on the resultant prototype.

Value engineering involves the generation of viable solutions and the selection of trade-offs between the part design, process plan and support design to achieve a cost-effective, defect-free and functionally acceptable prototype.

Communication among all parties pertain to the selection, justification and evaluation of a specific design, process plan and recommendation based on the ability to satisfy the given objective functions and constraints on resources.

A part design that focuses strictly on the aesthetic and functionality aspects will unnecessarily increase the building costs and delivery time. It may also affect the accuracy of the resultant parts and increase the cost for the secondary operations. An effective part designer must consider both the constraints and limitations faced by the RP process planner, fixture designer and value engineer.

On the other hand, a process plan which focuses strictly on the effective utilization of material and the RP machine may affect the aesthetic factors, functional requirements and costs required for the secondary operations. Similarly, an effective support design that concentrates on minimizing defects may affect the costs of production and secondary operations. A poor fixture design in turn may affect the accuracy, appearance and functionality of the resultant prototype.

An effective value engineer, on the other hand must consider all the available design/process options and determine the cost penalties involved if some of the functional, costs, time and quality constraints are violated. The role of value engineering is to develop as many alternatives as possible for the evaluation and customer advisory processes.

3.3 INTELLIGENT RAPID PROTOTYPING

The conventional rapid prototyping procedure involves a series of discrete independent functions: the creation of a CAD model, process planning to suit part accuracy and building, fixture design to suit part shape and building, verification and inspection, creation of an STL file, slicing of an STL model, part building, post curing or/and finishing operations. Of these, the CAD modeling, process planning, fixture design and verification and inspection are the most crucial and time-consuming functions required to produce an accurate and acceptable prototype.

From experience, the average time needed to design a part, develop a process plan and develop the required fixtures is of the order of three weeks for a product such as a telephone casing. This does not include the iterations required to satisfy the customer requirements of cost, accuracy, appearance, etc. This time could be reduced significantly if CAD modeling, process planning, fixture design and inspection were all integrated intelligently with knowledge-based systems with concurrent flow and control of information between each systems. Moreover, the iterations required to come up with an optimum design could also be performed concurrently and instantly. This is precisely the concept of an intelligent rapid prototyping (IRP) system.

The design and successful implementation of such a system involves the development of several intelligent knowledge-based systems (IKBS), each pursuing specific tasks, and then the development of an integrated environment to coordinate and control the activities and information and data exchange among these IKBS. This requires the adoption and combination of the best attributes of the following core technologies:

- feature-based modeling techniques
- object-oriented programming techniques
- intelligent knowledge-based technologies
- concurrent engineering
- distributed blackboard communication and control technologies.

The distributed blackboard approach is applicable to many complex tasks where several systems of expertise must interact closely to resolve a complex problem [5]. A typical example involves the integration of the design, planning, scheduling, monitoring, execution, feedback and tooling functions within a computer-integrated manufacturing (CIM) environment [6]. A successfully implemented intelligent rapid prototyping system within a distributed blackboard environment would achieve the following benefits:

- a higher level of productivity and greater utilization of resources
- a greater reduction in design, planning and fixturing time and costs
- savings in terms of speed, material usage and lead time
- a greater accuracy and quality of final products
- flexibility to react faster to a customer's demands
- better access and control of information at all levels
- an effective environment to facilitate cooperation between all personnel.

The major subsystems of the intelligent rapid prototyping (IRP) system comprise the following intelligent knowledge-based systems (IKBS):

- an intelligent feature-based RP part designer   (IFD) system
- a knowledge-based RP process planner   (KPP) system
- an expert RP fixturing support designer   (EFS) system
- a knowledge-based value engineering inspector   (KVE) system.

An intelligent feature-based RP part design (IFD) system provides fast and efficient part design and allows the product designer to specify both geometrical and non-geometrical attributes. Non-geometrical characteristics include the quality, aesthetic, surface finish, functionality, type of fits, geometrical tolerances and the permissible warpage or shrinkage of the resultant part.

A knowledge-based RP process planning (KPP) system aims to optimize the cost-effective production of high-accuracy parts. This is done through selecting the best orientation and positioning of part to reduce the laser processing time, material usage and set-up time.

An expert RP fixturing support design (EFS) system will provide the best design for the supports needed for any shape of the part with the aim of reducing the warpage, distortion and manual efforts needed for secondary operations.

A knowledge-based value engineering inspection (KVE) system will evaluate and arbitrate the differences among the above three sub-systems and will optimize the objective function within the constraints of the customer requirements and sub-system proposals.

3.4 CONFIGURATION OF THE IRP SYSTEM

The framework of a distributed blackboard-based intelligent rapid prototyping (IRP) system is shown in Fig. 3.1. The IRP system provides an integrated environment to coordinate the activities of the four knowledge sources, which are the intelligent knowledge-based systems of IFD, KPP, EFS and KVE. As shown in the figure, the main control elements of the IRP system comprise a blackboard, a scheduler, an event controller and an event activator. The main function of IRP is to monitor the exchange of both the given and deduced information and data among the four knowledge sources.

For example, the intelligent feature-based design system, IFD, will provide all the facilities to easily allow a human part designer to specify the necessary specifications. These specifications will then be broadcast concurrently, using the IRP system, to both the process planning (KPP) and fixturing (EFS) systems. The process planner, KPP, will evaluate the cost-effectiveness of a design and develop the process sequence of part orientation and positioning. The fixture designer, EFS, will determine the

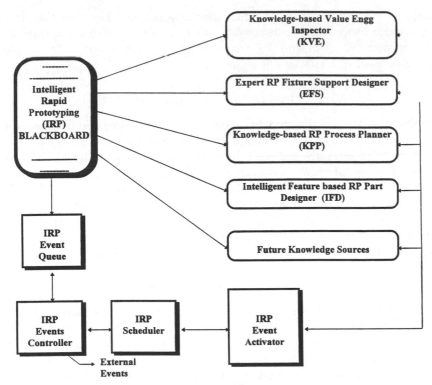

Fig. 3.1 Configuration of an intelligent rapid prototyping (IRP) system.

best support needed to hold the part concurrently. Both the resultant plan and the fixture design, along with the original part design, will be sent to the value engineering inspector, KVE, for verification. Whenever there is a conflict among the process planner, the fixture designer and the part designer, the KVE system will be required to select the best trade-off between the conflicting criteria.

The main technical advantage of grouping the four knowledge-based systems in a modular and independent form is that solutions can be developed incrementally. Multilevel problems can easily be represented using such a design framework. In addition, future functional knowledge sources can easily be added and integrated using the distributed-blackboard approach.

3.4.1 IRP blackboard

The IRP blackboard, as shown in the figure, is a shared and structured database that allows the four functional knowledge-based systems to interact anonymously. Each knowledge source has continual access to the state of the solutions on the IRP blackboard, and they can therefore contribute opportunistically. With such a strategy the appropriate information or data can be concertedly applied towards the rapid development of a concurrent solution.

3.4.2 IRP event controller

Problem-solving is directed by a flexible control component called the IRP event controller. The event controller is separate from the knowledge sources. The interface between the event controller and the knowledge sources is a set of events. Events are signals that are generated in response to an action within each knowledge source, such as the reaction towards a new piece of information. An event can also be an external happening which was previously defined as a significant occurrence, such as a mouse click or a time interrupt. The event controller is used for the creation, modification, updating or removal of the blackboard events.

3.4.3 IRP event activator

The IRP event activator as shown in the figure will consider a specific knowledge source as a candidate for activation, when an event occurs and matches with the trigger conditions. To verify whether a specific knowledge source should be activated, the IRP event activator will execute a predicate function, which has been previously defined. A predicate function is a prequalification rule that determines if sufficient conditions exist to start or interrupt an event. If the preconditions are sufficient, the event activator will start the relevant knowledge source by creating a knowledge source activator.

3.4.4 IRP event queue

The knowledge source activator is kept in a list of pending activities, called the IRP event queue. A rating can be assigned to each knowledge source activator to determine its position in the queue. The positional rating can also be automatically or manually defined using a predicate function for a specific knowledge source.

3.4.5 IRP scheduler

The main function of the IRP scheduler during the problem-solving process is to decide exactly what to do next, given a list of pending knowledge source activators, KSAs. It will activate the first KSA, followed by the next KSA. The problem-solving process continues until the event activator indicates that the scheduling cycle is to stop or interrupt until no more KSAs remain in the event queue. The scheduler in essence provides the mechanism to govern the states of events to be executed.

3.4.6 An example: exchange of information between IRP knowledge sources

In order to illustrate the exchange of information within the IRP blackboard node, IBBN, let us assume that a new fixture will be required for the RP part designed by the RP part designer.

Under such a condition, the IBBN, as shown in Fig. 3.1 will automatically be notified by the intelligent feature-based part designer (IFD). The IBBN in turn will create an IRP event for further action by the IRP event activator

The IRP event activator in turn will determine whether a predicate function has been defined to start the expert RP fixture designer (EFS). In addition, the IRP event activator will further establish whether sufficient information exists to make activation worthwhile.

The IRP event controller as shown in Fig. 3.1 will activate the expert RP fixture designer with a knowledge source activator, KSA, if sufficient conditions exists. The KSA is a list of pending events to be executed. These KSAs are placed in an IRP event queue.

The main function of the IRP event scheduler as shown in Fig. 3.1, during the information exchange process, is to decide exactly what to do next, given a list of KSAs. It will initiate the first KSA, which in turn will trigger the expert RP fixture designer. The function of the expert RP fixture designer when activated by the IRP event scheduler is to design the proper support considering various factors within its knowledge base and the constraints imposed by the part shape and size and the RP process.

The following sections describe the functions and activities of the four inter-related intelligent knowledge-based systems used in the IRP configuration.

3.5 INTELLIGENT FEATURE-BASED RP PART DESIGNER (IFD)

The basic function of the IFD is to provide an environment for the effective design and modeling of the intended RP part, taking into consideration both geometrical and non-geometrical attributes. Conventional techniques such as solid and surface modeling systems are slow and inefficient, particularly for the development of complex shaped artifacts. In addition, the facilities to represent other attributes, such as functionality, non-geometrical features and other engineering information, are insufficient.

In order to remove the deficiencies of the conventional CAD system, a feature-based design approach is adopted within the IFD. It consists of a parametric library of standardized engineering features such as slots, pockets and fillets. These features are developed using a commercial

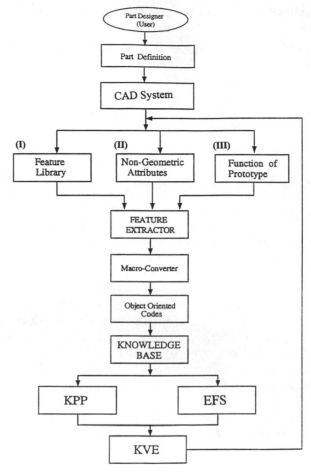

Fig. 3.2 Outline of intelligent feature-based RP part designer (IFD).

CAD software. The human part designer interacts directly with the IRP system through the IFD. Fig. 3.2 shows an overview of the IFD.

One of the main functions of the IFD is to encourage standardization of part design and to evaluate a design in terms of rapid-prototyping. The IFD will also generate detailed engineering drawings of the parts, together with all the relevant dimensions. The IFD system consists of several modules, and their functionalities are described below.

3.5.1 Part definition module

The part definition module allows a designer to define, select and position the geometrical features and to specify the functional attributes of a workpiece using a set of predefined elements from a CAD library. These features include derivatives of shapes, holes, slots, pockets, bosses, steps, ribs and other standardized machining and molding features. A user interface within the CAD system enables the designer to easily construct the shape of the object and to position the selected features with reference to a set of predefined datum.

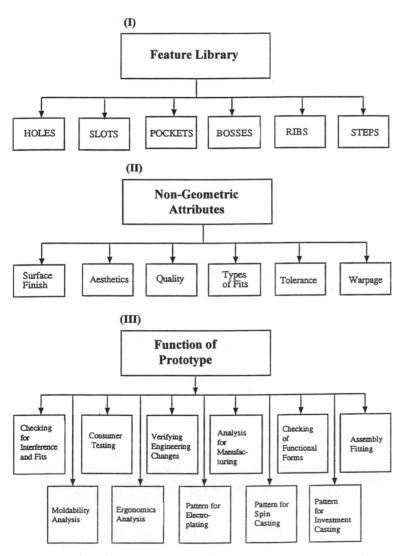

Fig. 3.3 Details of geometrical, non-geometrical, and functionality attributes within the IFD system of Fig. 3.2.

A unique feature of the part definition module is that it further allows the designer to specify the non-geometrical attributes and the functionality of the resultant artifact. Figure 3.3 shows the details of the geometrical, non-geometrical and functionality attributes within the part definition module. The non-geometrical attributes include requirements such as surface finish, aesthetic, shrinkage factors, tolerances, types of fits and permissible warpage. These functionality requirements provide a statement to specify the purpose and objective of the resultant RP prototype. Some of the main reasons needed by industry for the development of a prototype model include the following:

- interference, engineering change and functional form verification
- consumer testing and ergonomic analysis of shape, size, appeal and aesthetic
- pattern-making for spin casting, lost-wax investment casting, electroplating, arc spraying and mold-making
- ease of molding, manufacturing and assembly analysis.

3.5.2 Feature extraction module

Once the definition of the part and its associated characteristics have been fully defined, the feature extraction module will compile the part description into a list of macro-statements. These macro-statements describe the part, together with its associated features. The feature extraction process involves the consolidation of all the geometrical and non-geometrical features from the CAD system. The attributes that relate to the extracted features together with their geometrical relationships will serve as inputs to the macro converter, as described below.

3.5.3 Macro-converter module

The function of the macro-converter module is to translate the part macro-statements into an object-oriented format. All the geometrical features, together with the non-geometrical features and functionality requirements, are translated into a set of detailed object-oriented representation. The object-oriented codes are statements that define the relationships between the geometrical features and its functionality. These object codes will serve as the required input needed by the process planner (KPP) and the expert fixture designer (EFS) systems.

3.6 KNOWLEDGE-BASED RP PROCESS PLANNER (KPP)

The main function of the KPP is to develop a cost-effective process plan to build the RP parts which will conform to the given accuracy and quality requirements. This is achieved by the determination of a suitable orientation with the aim of reducing the laser time (as in SLA and SLS), material usage, set-up time and effort for the secondary operations such as polishing.

An overview of the KPP is shown in Fig. 3.4. It initially receives the part definition data and the functional specifications from the IFD in the form of an object code file. The non-geometrical specification also provides a guideline to assist the KPP to determine the purpose or function of the resultant RP part. The non-geometrical specifications also provide the necessary guidelines for the KPP to focus on a specific objective function. The next sections describe the sub-modules of the KPP.

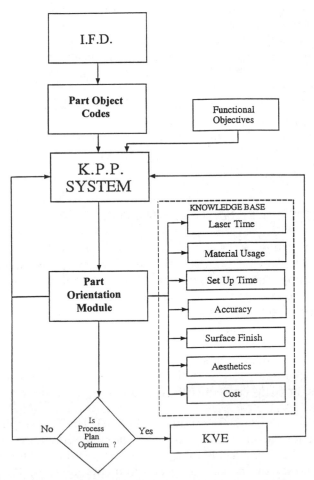

Fig. 3.4 Outline of knowledge-based RP process planner (KPP).

3.6.1 Part orientation module

The part orientation module will determine the most appropriate workpiece orientation to satisfy the objective functions of process planning, to minimize costs and time. It will initially establish if the CAD model is residing in the appropriate cartesian octant. The orientation module will also verify that the size of the final part will fit adequately on the RP machine building envelope. These are some of the hard constraints, and if they are contradicted the part orientation module will provide a warning message and advise the human part designer of such a violation.

The part orientation module contains rules to orientate a part to achieve any of the desired results, such as: minimize the height of the object as much as possible; or minimize the number of slanted surfaces to reduce the staircase appearance of the resultant parts, or maximize the number of smooth or aesthetically important surfaces.

For each specific part orientation, the KPP system will determine the material usage, set-up time, laser time and establish the overall cost associated with the RP process. The KPP will also predict the level of accuracy, surface finish and assign an aesthetic value to each chosen orientation. If the overall process plan is found to be suitable based on a specific objective function, then the process plan is forwarded to the knowledge-based value engineering inspection (KVE) system for value analysis.

3.7 EXPERT RP FIXTURING SUPPORT DESIGNER (EFS)

The main objective of the EFS is to develop the most suitable support design needed, to constrain the solidified layers and workpiece during the RP building process. Its principal aim is to minimize the amount of defects and the manual effort required for secondary operations. In many processes, an object is normally held by supports rather than being placed directly on the building platform. The main reasons for using these supports are to facilitate the separation of the main part from the platform without causing any damage, to constraint the overhanging layers during the building process, and to facilitate the drainage of the resin polymer quickly (as in SLA) to reduce post-processing time and potential air traps.

An overview of the EFS is shown in Fig. 3.5. The EFS system initially receives the RP part definition data from the IFD in the form of an object code file. Its key function is to select the most suitable support from a library to satisfy the functionality requirements given by the IFD. The sub-modules within the EFS system are described below.

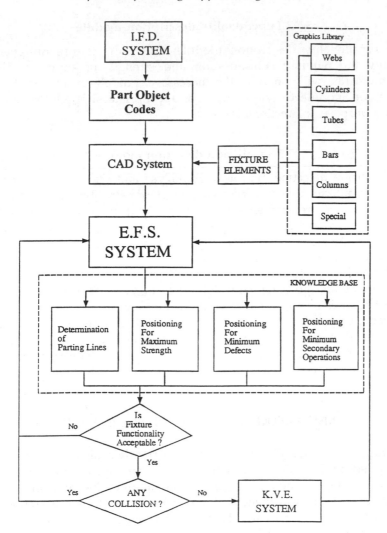

Fig. 3.5   Outline of RP fixture support designer (EFS).

3.7.1 Fixtures definition module

The function of the fixtures definition module is to select and position the supports using a set of predefined standard elements from a CAD library. These fixturing support in the CAD library include variations of different honeycombed bars and columns.

The fixture definition module contains design rules to assist the designer in the determination of the most suitable type of supports, spacing requirements and the effective placement of the selected elements. The number of supports should be minimized to reduce the processing time and the manual effort to remove them.

3.7.2 Functionality description module

The functionality description module allows the fixture designer to specify the required configuration and strength of supports. These requirements will provide the necessary guidelines to the fixture definition module to select the most appropriate supports to suit the stated purpose or functionality.

3.7.3 Collision detection module

The collision detection module will analyse and identify any potential collision, interference or excessive contact pressure between the work-piece and the recommended supports. It graphically displays the combined workpiece and supports the assembly as solid models. Potential collisions are detected by performing a boolean intersection operations within the CAD system.

The collision detection algorithm will ensure that excessive contact pressure points and potential weight distortion are eliminated. The areas of potential collision, actual amount of interference and pressure points are reported to the EFS rule-based module for re-adjustments. If further adjustments cannot be made due to the physical constraints, the EFS will re-select the next most appropriate support in an iterative manner to remove all the likely defects at the preliminary design stage.

3.8 KNOWLEDGE-BASED VALUE ENGINEER (KVE)

The function of the KVE system is to concurrently evaluate the proposals submitted by the IFD, KPP and EFS. It will attempt to satisfy both the given objective function of the design in relation to the constraints faced by the three systems. Whenever there is a conflict of interest among the three knowledge sources and the given objective functions, the KVE will attempt to resolve their differences. The main activities of the KVE system are summarized in Fig. 3.6.

To illustrate the interactions between the part designer, IFD, process planner, KPP, fixture designer, EFS and the value engineering system; KVE, the flow of information among these systems and the arbitration process are described in the next section.

3.9 INTERACTIONS BETWEEN THE IFD, KPP, EFS AND KVE

The IFD with the assistance of the part designer will initially create a model of the RP part. The geometrical and non-geometrical features of the objective function of the model are given by the part designer. The

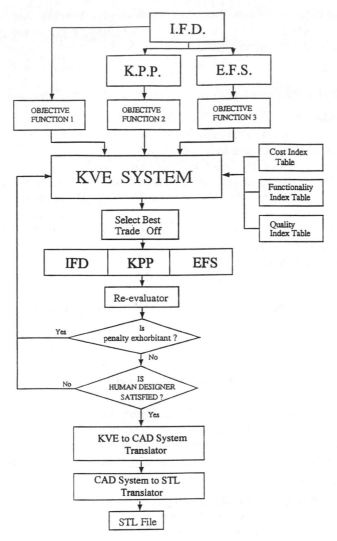

Fig. 3.6 Activities of knowledge-based value engineer (KVE).

IFD for instance, will attempt to develop a specific workpiece design to satisfy a particular objective function, as follows.

3.9.1 Objective function 1: maximize appeal and feel of the resultant part

This specific objective function is deemed to be the most important by the part designer. This initial workpiece design will be given to both the KPP and EFS simultaneously.

The KPP, when given the workpiece design without the objective functions given by the IFD or the EFS, will propose a process plan that will satisfy its own objective function, as follows.

3.9.2 Objective function 2: minimize machining time and cost

In satisfying its own goal, the KPP system may propose a specific workpiece orientation that will minimize machining time and cost. This recommended orientation in an attempt to minimize the machining time and cost may violate the objective function 1, which is given by the IFD (that is to maximize appeal and feel of the resultant part) or it may contradict the objective functions chosen by the EFS as describe below.

The EFS, when given the workpiece design, will select and position the necessary support to constrain the workpiece. In its attempt to design the support, the EFS system may propose a specific configuration with an objective function, as follows.

3.9.3 Objective function 3: minimize warpage and distortion

In satisfying its objectives, the EFS, after assessing the part shape, size and weight, will develop a specific support design that will minimize warpage and distortion. The fixture design developed by the EFS in its attempt to minimize warpage and distortion may also violate the objective functions given by the IFD and KPP.

The three solutions given independently by the IFD, KPP and EFS together with their key objective functions will serve as an input to the KVE. The KVE will attempt to assess each objective function and select a specific recommendation based on the best trade-off between the following criteria:

- the ability of the proposal to satisfy the functional requirements
- the ability to satisfy the qualitative requirements
- the ability to reduce costs and time.

As all these criteria cannot be satisfied simultaneously, the KVE will attempt to balance the independent criterion using a series of indices. The indices for each class of criterion are kept in the following lookup tables:

- the cost index table
- the functionality index table
- the quality index table.

Based on the relative strength of each composite index, the KVE will select one of the principal objective functions (say to minimize costs) to be the pivot for further re-analysis and redesign by the IFD and EFS.

The selected proposal and its objective function will be transmitted to the affected knowledge sources, namely the IFD and EFS. The affected systems which do not comply with the selected objective function, will make changes and re-adjust their original objective functions according to the intent of the KVE system.

The revised plan and fixture design, together with the revised objective functions and other penalties involved, will be resubmitted to the KVE system. The KVE system will re-evaluate the penalties involved. If the penalty involved is exorbitant in terms of costs, functionality and quality, the KVE will renominate a specific objective function as the new pivot to be considered by the affected knowledge sources.

This iterative process is repeated until a compromise between the costs, quality and functionality can be democratically reached. If a democratic solution cannot be amiably established, the KVE will pose all the potential solutions to the human designer for final arbitration. After the human designer has selected a specific proposal, the KVE system will merge and translate the recommended orientation and support design into a STL file which is suitable for the RP preprocessor.

3.10 APPLICATIONS AND DISCUSSIONS

Although the methodology of integrating several knowledge sources together using the distributed blackboard approach has been available for some time, the application of this technique to rapid prototyping systems is a recent phenomenon. Work has recently been done in applying the methodology to two specific rapid prototyping processes, the stereolithography and the fused deposition modeling [7,8]. In these applications, though the generalized approach is the same, the specific requirements and characteristics of SLA and FDM have been highlighted and included in the development of the knowledge sources and their integration with the distributed blackboard architecture. The blackboard architecture has also been effectively applied to computer-integrated manufacturing, process control, computer vision, planning and scheduling, and shop floor scheduling, where various cooperating expert systems have to work together to satisfy target demands [9].

The distributed blackboard framework for intelligent rapid prototyping provides an effective platform for the integration, control and exchange of information concurrently through its cooperating knowledge sources. Complex relationships between various functions are easily defined using the IRP framework. These relationships include functions related to the customer specifications, CAD modeling, part orientation, process planning, fixture supports configuration and value analysis and inspection. With the implementation of IRP environment, the necessary and accurate information can thus be made immediately available for fast and economical production of RP prototypes.

Another important advantage is that the necessary data can be rapidly generated for the designer as well as for the customer of the RP part. These include the detailed part drawing and design information, the bill of materials, the cost estimation, the level of expected accuracy and quality, the listing of standardized components, the material volume, and the fixturing requirement information. The ability to respond immediately to a customer request is currently one of the major problems in a market-driven manufacturing environment. This is largely attributed to the inability of the human engineer to capture and retain all the necessary information to satisfy customer expectations.

By integrating a set of knowledge sources into the blackboard architecture, the IRP system will provide effective conflict resolution strategies peculiar to the nature of any of the rapid prototyping technologies. This is largely due to the fact that in such an environment, precise information can be channelled in the right direction at the right time. This also eliminates the need for large and centralized databases which are often necessary with conventional manufacturing and prototyping systems.

3.11 CONCLUSIONS

The intelligent rapid prototyping framework presented in this chapter provides a methodology to improve the efficiency and capability of a rapid prototyping system through the application of distributed blackboard control technology. Such a framework facilitates not only the marriage of cooperating knowledge sources but also the true integration of all personnel in the company to develop the most attractive proposal, effective design, execution plans, and the best products. This collaborative effort will then enhance the economic performance of an enterprise.

Though the IRP system includes and integrates four major intelligent knowledge-based systems (IKBSs) related to the prefabrication activities of the rapid prototyping process, the domain of the IRP implementation can be enhanced by the inclusion of other IKBSs related to other aspects of the process. This may include STL preprocessing and verification, optimization of faceted approximation of the model, slicing strategies and considerations related to post-curing and secondary operations.

REFERENCES

[1] Rapid prototyping report (1995) CAD/CAM Publishing Inc.
[2] Hull, C. *et al.* (1995) Rapid prototyping: current technology and future potential. *Rapid Prototyping Journal*, 1(1), 11–19.

[3] Aubin, R. F. (1994) *A world wide assessment of rapid prototyping technologies.* UTRC Report No. 94/13, United Technologies Research Centre, East Hartford, CT, USA.

[4] Masood, S. H. *et al.* (1994) Evaluation of CNC and SLA as prototyping techniques. *Proceedings of the IMS International Conference on Rapid Product Development*, Stuttgart, Germany, 349–360.

[5] Jagannathan, V. *et al.* (1989) *Blackboard architecture and applications*, Academic Press, Boston, MA.

[6] Lim, B. S. (1992) CIMIDES: A computer integrated manufacturing information and data exchange system, *International Journal of CIM*, **5**(4/5), 240–254.

[7] Masood, S. H. and Lim, B. S. (1995) Concurrent intelligent rapid prototyping environment. *Journal of Intelligent Manufacturing*, **6**(5), 291–310.

[8] Masood, S. H. (1996) Intelligent rapid prototyping with fused deposition modelling. *Rapid Prototyping Journal*, **2**(1), 24–33.

[9] Lander, S. E. *et al.* (1991) Knowledge-based conflict resolution for cooperation among expert agents. *Computer-Aided Cooperative Product Development*, Sriram, D. *et al.* (eds), Springer-Verlag, Berlin, 253–268.

CHAPTER FOUR

Investigation on surface features of rapid prototyping parts

L. Felloni, A. Gatto, R. Ippolito and L. Iuliano

4.1 INTRODUCTION

The creation of a prototype is an essential step in the elaboration of a new product.

In the current situation where product lifetimes are continually decreasing and design complexity increasing, the rapid creation of models (rapid prototyping) offers a possibility to reduce the 'time to market' and react faster to the demands of the market. The prototype produced by RP technique is currently progressing from a style model [1] to a working part [2], and hence the roughness and the surface morphology of rapid prototyping parts are crucial features in many cases, and sometimes they determine the fields of application.

The surface of a manufactured part generally has properties and a behaviour that are different from those of its bulk. This is due to the mechanical, physical, thermal and chemical effects induced by the manufacturing process [3]. Non-uniform surface deformation or severe temperature gradients during manufacturing usually cause residual stresses in the surface structure. The bulk material influences the component's overall mechanical properties, whereas the component's surface finish influences not only the dimensional accuracy of machined parts, but also several important properties and characteristics of the manufactured part:

- the appearance and geometric features of the part and their role in painting, coating, and adhesive bonding
- the initiation of cracks because of surface defects could lead to weakening and premature failure of the part

- the thermal conductivity of contacting bodies; rough surfaces have higher thermal resistance than smooth ones.

For these reasons the surface finish is often corrected by means of manual operations but this causes a loss in precision and an increase in production times and costs, and hence a factor that ties the result to the ability of the operator is introduced. The final result, however, is influenced by the material chosen, and hence the operating parameter values. It is therefore still greatly dependent on the skilfulness of the operator [4].

Surface integrity describes the topological and geometric aspects, properties and characteristics of surfaces. The defects produced during component manufacturing can be responsible for a lack of surface integrity caused by:

- defects in the original material
- the method by which the surface is produced
- the lack of proper control of the process parameters.

Flaws or defects, such as scratches, cracks, holes, depressions, seams or tears, are random irregularities.

Roughness may be superimposed on waviness which is a recurrent deviation from a flat surface. Roughness consists of closely spaced, irregular deviations on a smaller scale than waviness. Waviness may be caused by deflections of the workpiece or warping from forces or temperature in the system during the manufacturing process. It is measured and described in terms of the space between adjacent crests of the waves (waviness width) and the height between the crests and valleys of the waves (waviness height).

4.2 SEM OBSERVATION OF THE RP PARTS

The chapter deals with the manufacture and testing of user accuracy parts [5, 6] produced by different RP techniques. The RP techniques used were: stereolithograpy apparatus (using two different machines); solid ground curing; selective laser sintering (using polycarbonate and nylon); fused deposition modeling (using machinable wax, polyamide and polyolefin); and laminate object manufacturing.

The surfaces and the cross-sections of the RP parts were sputter-coated with a gold layer using the following parameters:

- current = 20 mA
- discharge time = 180 s
- pressure = 2×10^{-2} bar.

and they were then observed at the SEM using both secondary and back-scattered electrons.

4.2.1 Stereolithograpy apparatus

The surface of the piece made with the SLA (3D Systems) technique is regular and it is characterized by growth steps due to energy absorption when the laser beam crosses the liquid photopolymer (Figs 4.1 and 4.2).

An incorrect post-process can cause residual stresses in the surface structure, which may be responsible for the delamination of layers as shown in Fig. 4.3. Figure 4.4 shows a photoelasticity image of two circular samples with different growth direction. The presence of a stressed layer in only one sample confirms that the layer's growth direction influences the surface residual stresses. In Fig. 4.5 the inner structure of pattern to be used directly in the casting process is visible: this Quick Cast part was built using a 3D Systems machine. With the Quick Cast part the turnaround time is half that of traditional methods of casting.

Figure 4.6 shows the surface of a SLA part produced with an EOS machine. Small cavities and irregularities are present.

4.2.2 Solid ground curing

The surface of the piece produced with the SGC is more irregular than the SLA part (Fig. 4.7). Moreover, its cross-section shows the presence of

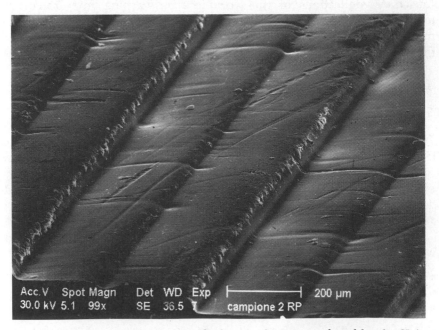

Fig. 4.1 SEM micrograph showing the layers of a part produced by the SLA technique (3D Systems).

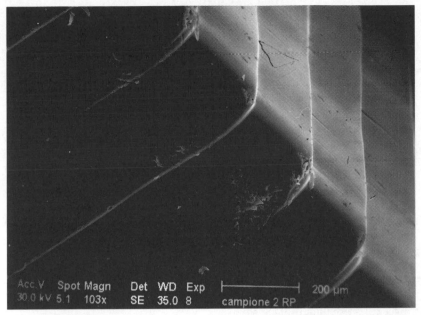

Fig. 4.2 SEM micrograph of a corner of a part produced by the SLA technique (3D Systems).

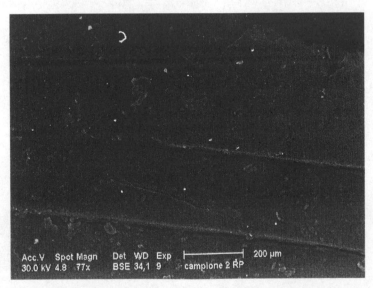

Fig. 4.3 Decohesion of the outer layers due to residual stresses (SLA technique, 3D Systems).

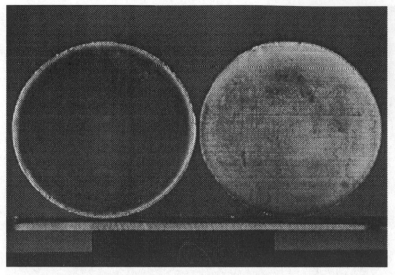

Fig. 4.4 Photoelasticity image of two circular samples with different growth direction; the white outer circular layer indicates the presence of residual stresses (SLA technique, 3D Systems).

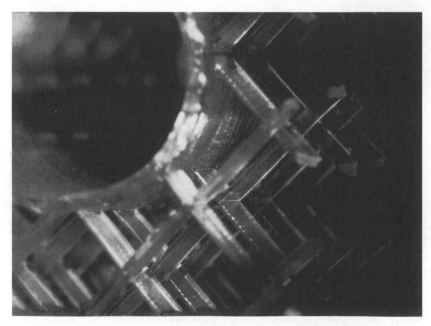

Fig. 4.5 Inner structure of Quick Casting part (SLA, 3D Systems).

Fig. 4.6 SEM micrograph showing the layers of a part produced by the SLA technique (EOS).

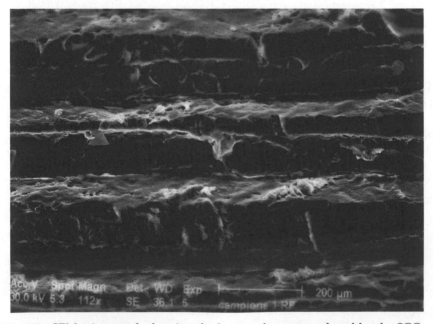

Fig. 4.7 SEM micrograph showing the layers of a part produced by the SGC technique (Cubital).

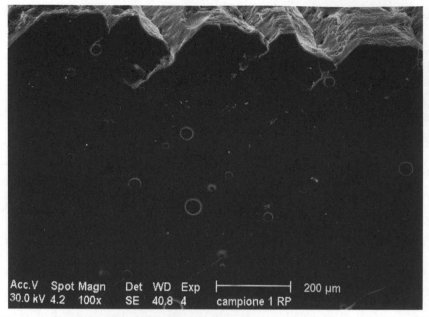

Acc.V Spot Magn   Det WD Exp ├───────────┤ 200 µm
30.0 kV 4.2 100x   SE   40,8 4   campione 1 RP

Fig. 4.8 Cross-section surface of an SGC part.

spherical microcavities at the interface between two layers but they are situated in the layers polymerized during the previous pass (Fig. 4.8).

4.2.3 Selective laser sintering

The SLS technology allows the use of a variety of materials. In these observations nylon and polycarbonate were used.

The irregular and porous structure of the surfaces of the SLS parts clearly demonstrate that it is only the portion of the particle actually exposed to the laser beam which melts and bonds to that next to it.

This irregularity is more marked in the specimen made with nylon powder (Fig. 4.9) probably because the intensity of the beam was reduced to minimize the distortion that this technique may cause when used to produce very thick parts. Figures 4.10 and 4.11 show the aspects of the surface and of a corner of a part made with polycarbonate.

4.2.4 Fused deposition modeling

Currently three materials are used: polyamide, polyolefin and investment casting wax, and all of them were studied.

In Figure 4.12 the painted surface of a polyamide part is shown; the visible crack is evidence of the imperfect junction of one layer with the

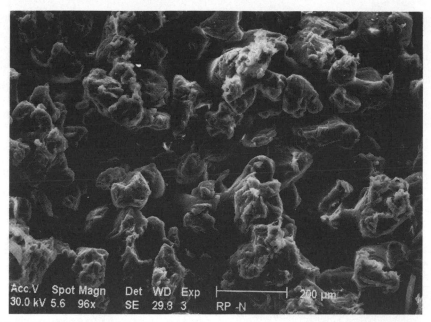

Fig. 4.9 SEM micrograph showing the surface of a part produced by the SLS technique (nylon).

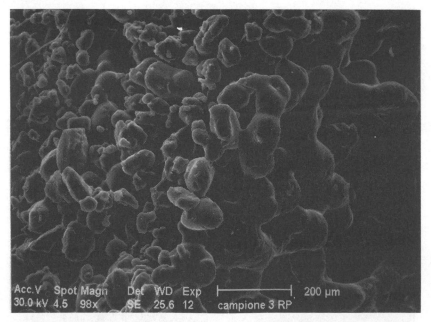

Fig. 4.10 The porous structure of a corner of a polycarbonate part produced by the SLS technique; the two surfaces appear different because only a portion of the particle exposed to the laser beam melts.

Surface features of rapid prototyping parts

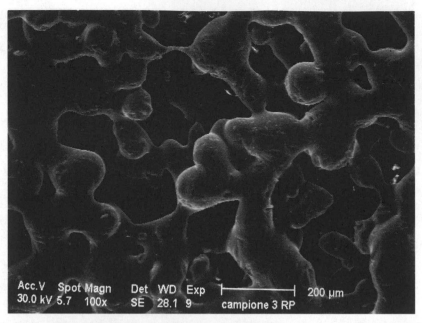

Fig. 4.11 SEM micrograph showing the surface of a part produced by the SLS technique (polycarbonate).

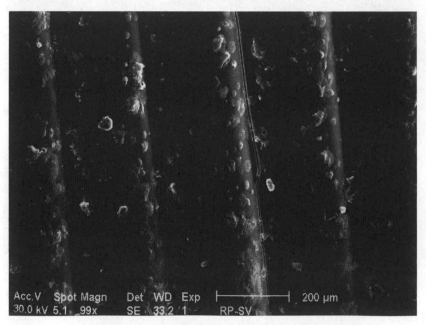

Fig. 4.12 Painted surface of a polyamide part produced by the FDM technique.

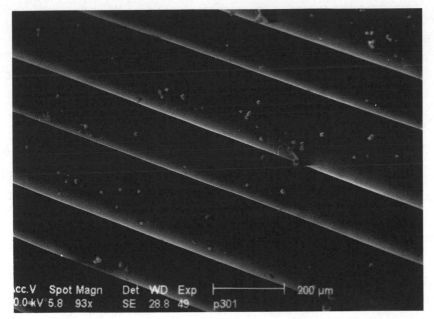

Fig. 4.13 Surface of a polyamide part produced by the FDM technique.

next, probably due to the inappropriate machining parameter values [4].

The growth layers of the polyolefin part are less regular and sharp than the polyamide and wax ones (Figs 4.13–4.15). On observing the lateral surface, the growth layers of the wax model are present and they display rather regular steps on which roughness depends (Fig. 4.15).

The pieces produced by the FDM technique show that continuity is not always achieved between adjacent layers.

In the model produced by fused deposition modeling filling problems may be present when the wall thickness of the part is high. Indeed, cavities may be visible in the filling within the inner zone bounded by the perimeter of the model for each growth surface, as shown in Fig. 4.16.

4.2.5 Laminate object maufacturing

The parts produced by the LOM technique need a post-process treatment. To guarantee the model durability and to eliminate environmental effects, the surface is manually finished and painted. To observe the growth layers, the part was sectioned and dry metallographically prepared. In Fig. 4.17 the growth layers are shown.

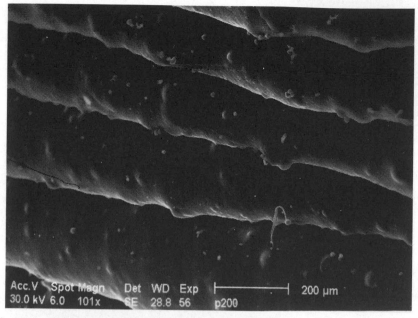

Fig. 4.14 SEM micrograph showing the surface of a part produced by the FDM technique (polyolefin).

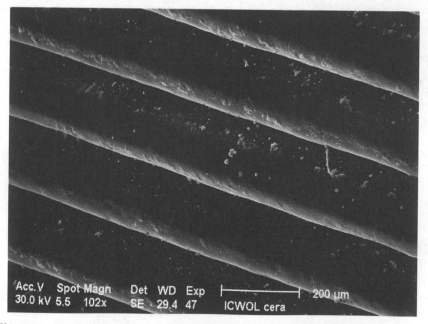

Fig. 4.15 Growth layers of investment casting wax model produced by the FDM technique.

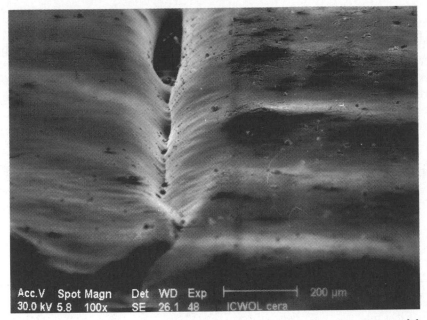

Fig. 4.16 Cavity due to filling problems in an investment casting wax model.

Fig. 4.17 Growth layers of a model produced by the LOM technique observed by OM, 200X.

4.3 SURFACE ROUGHNESS

The surface roughness Ra was measured in parallel and perpendicularly to the growth layer's direction to give a better idea of the influence of each technique, because it will be recalled that their mode of generation renders these surfaces greatly anisotropic. This parameter was measured, on flat surfaces, with a Hommel T1000 Stylus instrument using a sampling length of 4.8 mm.

Table 4.1 Comparison of the results obtained by the RP techniques in terms of mean surface roughness values measured on flat surfaces

Technique	Material	Ra (μm)	
		parallel	perpendicular
SLA (3D Systems)	SL 5170 epoxy resin	3.7	1.7
SLA (EOS)	epoxy resin	1.8	1.1
SLA(Quick Cast)	SL 5170 epoxy resin	2.8	1.6
SGC	G-5661 polyester	9.8	5.0
LOM	paper	4.0	2.4
FDM	P301 polyamide	11.0	1.2
	P200 polyolefin	11.2	5.0
SLS	polycarbonate	16.6	15.3
	nylon	14.8	1.1

Differences appear in the mean roughness values reported in Table 4.1 and in Fig. 4.18. It should be noted, however, that the low value for the LOM process may be attributed to the fact that the part is made of paper and must be painted to ensure that its dimensional stability is not influenced by the ambient humidity.

The worst surface roughness is displayed by the SLS parts; manual finishing would be necessary to improve the surface quality. Better results are obtained by SLA process also using the Quick Cast building mode, which means that low surface roughness is an intrinsic feature of this process and depends only on the layer thickness.

On the FDM wax part it is obviously impossible to measure the roughness with a contact instrument, therefore the Ra values are not included in the Table 4.1.

To better understand the performance of the RP techniques in terms of surface roughness, this parameter was measured along a non-flat surface,

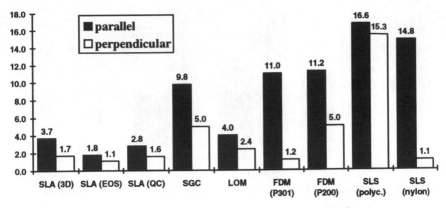

Fig. 4.18 Comparison of the roughness measurement on RP flat parts.

spherical or cylindrical. In this case it is possible to note the step effect due to the slicing process. Moreover, these surfaces are typical of moulded items, such as the plastic parts of a car or a small object used at home and in the office.

A comparison is made among the techniques stereolitography, solid ground curing and selective laser sintering. The surface roughness was measured with a Talysurf instrument using a sampling length of 15 mm.

Table 4.2 and Fig. 4.19 report results that demostrate that there is a marked drop in the surface roughness of non-flat surfaces respect to the flat ones.

The roughness of all the parts was measured 3 months after manufacture to assess the stability over time, and no marked differences were found.

Table 4.2 Comparison of the roughness measurement carried out on non flat surface

Technique	Material	Ra (μm)	
		sphere	*cylinder*
SLA (3D Systems)	SL 5170 epoxy resin	25	39
SGC	G-5661 polyester	26	40
SLS	nylon	18	19

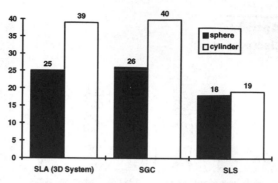

Fig. 4.19   Comparison of the roughness measurements on RP no flat parts.

On the basis of the roughness measurements carried out on the RP parts the following observation can be made:

- the *Ra* of the flat part is good and is comparable with the material removal processes
- the roughness of the non-flat parts at the moment is not good, and manual finishing of the parts is necessary for their rapid tooling employment. In the future, with the evolution of software, the adaptive slicing will probably help to resolve the actual problem.

REFERENCES

[1] Kruth, J. P. (1991) Material incress manufacturing by rapid prototyping techniques. *Annals of the CIRP,* **40**(2), 603–614.
[2] Bjorke, O. (1991) How to make stereolithography into a practical tool for tool production. *Annals of the CIRP,* **40**(1), 175–178.
[3] Kalpajkjian, S. (1991) *Manufacturing process for engineering materials.* Addison-Wesley Publishing Company, 1991, 159–162.
[4] Ippolito, R., Iuliano, L., Gatto, A. (1995) Benchmarking of rapid prototyping techniques in terms of dimensional accuracy and surface finish. *Annals of the CIRP,* **44**(1), 157–160.
[5] Gargiulo, E. P. (1992) Stereolithography process accuracy, user experience. *Proceedings of the First European Conference on Rapid Prototyping,* Nottingham, U.K., 187–201.
[6] Jacobs, P. F. (1992) *Rapid prototyping and manufacturing, fundamental of stereolithography.* SME, Dearbon, USA.
[7] Jacobs, P. F. (1995) *Quick Cast 1.1 and rapid tooling. Proceedings of the Fourth European Conference on Rapid Prototyping,* Belgirate, Italy, 1–26.

CHAPTER FIVE

Rapid prototyping using fiber-coupled high-power laser diodes

Jian (John) Dong, Tariq Manzur and Chandra S. Roychoudhuri

5.1 INTRODUCTION

In today's global market place, quickly bringing a high-quality new product to the market has become a critical factor in a company's success. Rapid prototyping (RP) or solid freeform fabrication (SFF), a group of emerging technologies for the production of mechanical components without part-specific tooling, provides great potential for rapid product realization.

SFF aims at rapid prototyping of physical objects directly from computer-based descriptions of the geometry of objects such as solid models. The rapid prototyping industry is one of the fastest growing industries in recent years. In the late 1980s, 3D Systems Inc. (CA) was the only company involved in manufacturing rapid prototyping machines. Currently, more than a dozen companies are making rapid prototyping machines with various available techniques. The techniques are stereolithography (3D Systems, CA), selective laser sintering (DTM, TX), solid ground curing (Cubital, Israel), fused deposition modeling (Stratasys, MN), laminated objects modeling (Helisys, MI) and 3D printing (MIT, MA), etc. The overall RP machine sales have grown by over 400% in five years 1988–93, Nakagawa, 1994). It is expected that RP industries will keep their fast-growing pace if the improvement can be made in any of the following areas.

- Parts from metal/ceramic powder: with the current rapid prototyping technologies, the majority of prototypes can only be made from wax, plastics, nylon and polycarbonate materials. Industries greatly wish to extend current techniques or new techniques to produce prototypes directly from metal and ceramic material.

- Cost of laser devices in RP machines: all current RP machines and solid freeform fabrication researches are using high-power lasers to melt or solidify materials to form a prototype. Excimer, CO_2 and Nd-YAG are the commonly used lasers. Although these lasers can produce high power, they are expensive, inefficient, bulky and difficult to control.
- Mechanical quality of prototypes: most prototype parts made from the current RP machines and/or research has very low structural strength and integrity. The quality of a prototype depends heavily on the CAD data used to control laser beams, the materials and the sintering processes.

Researchers at the University of Texas at Austin and DTM Corporation (Crawford, 1994; Prabhu and Bourell, Glazer and Barlow, 1994; Forderhase and Corden, 1994; Zong, 1992) have patented their selective laser sintering (SLS) technology for the SFF of plastic, nylon and polycarbonate. Now they are trying to apply the SLS technology to metals. One of their techniques is to coat metal particles with a binder. The binder is selectively cured with a CO_2 or Nd: YAG laser, and the part is later fired to burn out the binder and make the part more dense. Other research in the area of solid freeform fabrication of metal/ceramic powder is 3D printing, which is being conducted at MIT (Yoo *et al.*, 1994).

In this chapter we demonstrate the way to sinter metal powders without any binder by using a laser diode (LD). The main reason for using a laser diode is that it will be very inexpensive (less than $10 W^{-1}$ in the future, compared with Excimer, Nd-YAG and CO_2 at more than $80 W^{-1}$. Cost reduction will come from the ability to batch-process thousands of laser diodes like semiconductor computer chips. In addition, laser diodes are very compact, fiber-deliverable, efficient, and controllable to any desirable pulse rate. A compact and inexpensive desk-top sintering machine could be developed from these in the near future.

5.2 THE DESIGN OF EXPERIMENTATION APPARATUS

The apparatus used for the experiment is shown schematically in Fig. 5.1. Figure 5.2 shows a photograph of the set-up. The apparatus includes a power system (diode laser), a powder feeding system, an oxidation prevention system and a laser scan control system.

5.2.1 Laser system design

In the world of material processing, Excimer, CO_2 and Nd-YAG lasers are used most widely in the area of welding, cutting, drilling, surface hardening, heat treatment, shaping, forming and other potential materials processing. However, these lasers are at their mature limits; they are

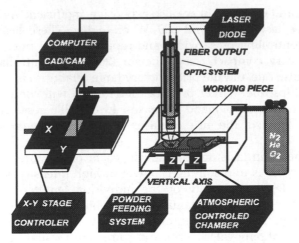

Fig. 5.1 Schematic diagram of laser sintering systems including a laser diode, an optical systems, a computer-controlled X–Y stage and a chamber.

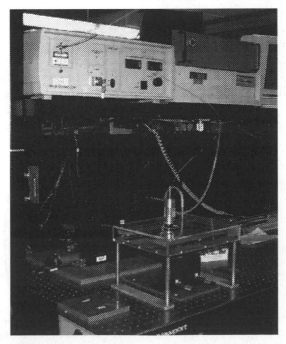

Fig. 5.2 Experimental set-up of laser sintering systems.

high-power and reliable, but are also expensive, inefficient and bulky. The cost of these lasers is around $80/W or higher. These lasers are also limited in controlling pulse width and repetition rates. In contrast, laser diodes are very compact, very efficient (20–50%), controllable to any desirable pulse rate and the price in very large quantities is ≤ $10 W. Like computer chips, the market price of diode chips will continue to come down as the volume demand increases. In our SFF research, we have exclusively studied the use of laser diodes as power sources for sintering.

One of the main tasks for the development of the direct laser diode sintering process is to obtain high power at high intensity. For an edge-emitting diode, the intensity of the individual emitter is limited to $1–10\,MW/cm^{-2}$ by the facet damage (or reliability) requirement. So high power at high intensity requires packing the diodes as densely as possible. As heat removal is a critical requirement for high reliability, it is better to transport the optical power by fibers from a thinly packed diode array and densely pack the fiber output ends to a remote site. The highest average output intensity requires the highest possible packing density of the fiber ends with thin cladding. However, the more important factor is the ability to couple the diode power very efficiently while preserving the emitter brightness. It is important to reduce the high emitter divergence by micro-optics and use fibers of closely matching cross-section and NA for the highest obtainable brightness.

With modular design concepts, multiplex (gather) high power from a large number of incoherent laser diode (LD) arrays is designed as the power system for the sintering process. During the diode laser system design, two critical technical issues have been studied: reduction of the beam divergence, or on increase in the brightness of the LDs; and efficiently collecting power from a large array of incoherent diodes.

There are many possible design approaches using individual diodes, 1D diode arrays, 2D diode stacks, coherent vertical cavity surface

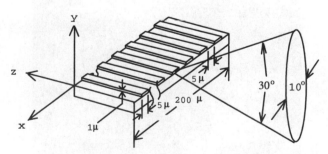

Fig. 5.3 Schematic diagram of a 1 W densely packed cluster diode stripes. The far-field beam divergence is elliptical (30° × 10° due to the rectangular emitting facet of each stripe.

emitting laser arrays, unstable cavity diode lasers, flared amplifiers, etc. The output beams are combined together through various combinations of fibers and lenses. These have been summarized in Chen, Roychoudhury and Banas (1994).

The particular laser systems used in this report consist of 31 fiber-coupled laser diodes. Thirty of these diodes, with each emitting 1 W, 980 nm wavelength laser beam, are individually coupled to 200 m core diameter multimode fibers. The 31st is a 10 mw visible diode for optical alignment. These 31 fibers are then closely packed and optically imaged onto a single 600 m core glass IR coated fiber. This 600 m core fiber becomes the flexible power delivery end, which is imaged with the appropriate magnification and demagnification on the metal powder for sintering

5.2.2 Powder feeding system

SFF using laser sintering generates 3D parts by multiple layers of powder sintered in a layer-by-layer manner. The powder feed mechanism will affect the productivity and quality of the sintered product. In this experiment a piston feeding system is designed with which powder can be fed manually. Automatic powder feeding will be incorporated into future system. Figure 5.4 is a sketch of the powder feeding system.

This system includes two pistons A and B and a roller. The right side (B) piston holds all the powder needed for a prototype, and left side piston (A) is the work area to perform sintering. To add a layer of powder, A is raised to the desired height, and B is lowered by the same height. Both piston movements can be controlled manually or by a computer through two stepper motors. The roller will push the powder from the area of B and push compacted powder hydrostatically into the area of A

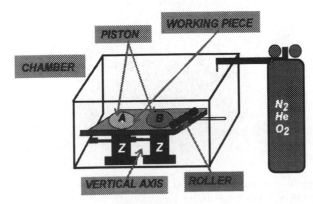

Fig. 5.4 Sketches of the powder feeding system; A and B are two pistons, the roller moves the powder from B to A.

before it is sintered. After one layer has been sintered, a new layer of fresh powder is laid down on the top of the sintered one. The process is repeated layer-by-layer to construct a 3D structure. To increase the structural strength, we have used static pressure to compress the powder layer-by-layer by a roller that moves the powder from B to A.

5.2.3 Oxidation prevention system

Oxidation is a major problem in the laser sintering process, because oxides often cause very loose structures and bonds between layers. To prevent oxidation, we have filled the sintering chamber with inert argon (Ar) gas.

5.2.4 Laser scan control

Laser scan control consists of two portions of CAD data generation, and a mechanical $X-Y$ slideway that holds and moves a laser scanner based on CAD data. Figure 5.5 is a sketch of the laser scan control system. The

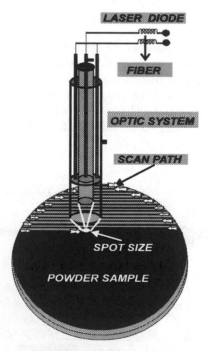

Fig. 5.5 Arrow indicates the laser scanning path controlled by a computer; a high-power (25 W, CW) laser diode (wavelength 950 nm) has a spot size of 2 mm and each scan overlaps by 455 the previous scan; the scan speed is 2.12 mm/s, the step size = 0.9 mm.

movement of the X–Y slideway is controlled by data generated by a computer through two step motors.

5.2.5 CAD data transferring

Most rapid prototyping processes produce parts on a layer-by-layer basis. The first step of the rapid prototyping is to slice the geometric description of the part (CAD data) into layers. The slicing operation generates the contours of the part for each layer. Each layer can then be further divided into many scanning paths. Three scanning paths are currently being studied. They are: raster scanning in a single direction (unidirectional scanning), directional scanning and contour scanning. Most current RP processes use raster scanning in a single direction to build each layer of the part. Unidirectional scanning will result in a large number of very short scanning vectors, and this requires a very precise and sensitive motor to control laser beam. Directional scanning results in a smaller number of longer vectors. Longer vectors reduce the errors associated with laser toggling transients and repositioning of the laser beam, resulting in higher part accuracy. Contour scanning of the layer boundaries is expected to generate a better surface quality (Crawford, 1993; Dong, 1996; Dong *et al.*, 1996).

The form of the geometric description of the mechanical parts to be produced by the sintering process significantly affects the accuracy of the final part. Most current technology consists of tessellating the surfaces of the geometric model into a mesh of non-overlapping triangular facets. The resulting geometry is transmitted in a standard file format, the so-called STL file format, developed by 3D Systems Inc. (1988). This format has been adopted by many CAD vendors and is readily available.

In our preliminary research, the control data are not from a CAD system; instead, the data are generated with user-written programs. For a cylindrical part (Fig. 5.5), a CAD system or a user program will first slice the cylinder into a number of disks, and then the laser will scan and sinter the powders path-by-path and layer-by-layer according to the program.

For more complicated parts, a solid modeling package I-DEAS (a product of SDRC) is available to generate STL files. A solid model must first be created, and then used to generate the STL file.

Because the generation of STL file consists of a tessellation process, the tessellation operation itself introduces errors in the model (Crawford, 1993; Dong, 1996). To overcome these problems, a new approach has been taken. In this approach 3D laser scanning path data for laser sintering are generated into the standard IGES and PDES/STEP file formats from non-uniform rational B-splines (NURBS) formats. The advantage of using the NURBS data instead of the commonly used triangular data is that it requires fewer data conversions for most standard analytical shapes (such as lines, conic, circles, planes and quadratic surfaces). The transfer of

CAD file data to a laser scan system is an important first step for rapid prototyping. This technique is expected to significantly affect the output of rapid prototyping with functional parts of improved tolerance and surface finish capabilities. The details of this research will be presented in another paper.

5.3 EXPERIMENTS AND RESULTS

Many factors will affect the quality of a sintered prototype. Based on a literature review and our experience, we have identified eight controllable factors that are considered critical to the direct laser sintering processes. These are:

(a) powder/size
(b) laser power
(c) fluence
(d) laser wave length
(e) laser spot size
(f) scan speed
(g) step size
(h) radius of a cylinder
(i) layer thickness
(j) process environment
(k) pressure on powder/binding powder
(l) scan pattern

Based on these controllable factors, we have designed numerous experiments, as follows.

5.3.1 Experiment on oxidation prevention

As already mentioned, the prevention of oxidation during direct sintering processes is one of the major challenges in rapid prototyping metal parts. Oxides cause very loose structures and bonds between layers. One way to prevent oxide is to use a chamber filled with an inert gas, such as Ar, He, N_2 and their mixture. The metal powder will be housed inside the chamber during sintering. However, although the atmosphere is free of oxide, enough residual oxygen in the powder will cause oxide formation. To solve the problem, we have used the pretreatment approach.

In the pretreatment approach, mix nano-composites with $CaO + H_2O$ and C and then keep it for at least 30 minutes under a flowing inert gas atmosphere or forming gas. Before sintering the powders are kept on a heated stage (100–300°C) for around 15 min to reduce oxygen from the powders ($CaCO_3 + H$). After the process, the powder is sintered in an inert gas chamber. The sintered parts are then washed in water to get rid of the $CaCO_3$.

5.3.2 Experiment on the pressed powder

One of the big challenges in the laser sintering of metal parts is to achieve compactness. Most laser sintered metal parts have very low densities (Badrinarayan and Barlow, 1992). In this research, we tested the influence of pressure added to powder each time a layer of powder was fed to a sintered part.

In the two experiments, the control parameters are:

(a) powder/size: iron 325 mesh, 99.99% (Alpha Products)
(b) laser power: 15 W (max. 25 W), CW diode laser (Applied Optronics Co.)
(c) fluence $F = 4 \times 10^2$ J/cm^2
(d) laser wave length $I = 980$ nm
(e) laser spot size 2–3 mm
(f) scan speed 2.12 mm/s (25.4 mm/s = 6000 counts/s)
(g) step size 0.9 mm
(h) radius of the cylinder 25.4 mm
(i) layer thickness 0.5 mm
(j) process environment: (1) air, (2) inert gas (Ar), (3) inert gas + pretreatment
(k) pressure on powder/binding powder: (a) no pressure added, (b) manually press powder each time before sintering a layer
(l) scan pattern: raster scanning

5.3.3 Experiment results and comparisons

Samples and microstructures

Figure 5.6 illustrates sintering in the situation of adding external pressure in an air atmosphere. The samples without adding pressure in air and/or argon atmosphere are the same as the one illustrated in Fig. 5.6. However, their microstructures are very different. Figures 5.7–5.9 show their microstructures.

Micro-hardness measurements of sintered samples

Microstructure evaluation

From the above figures it is clear that the external pressure during sintering reduces the porosity and increases the compactness of the finished parts. The sintered sample in air is oxidized and it forms large voids, compared with the sintered sample in inert (Ar) atmosphere.

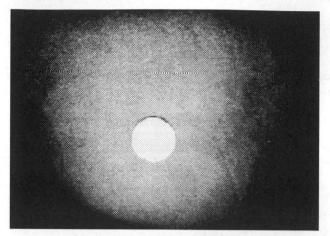

Fig. 5.6   A diode laser sintered sample (manually pressed powder each time a layer of powder is fed to the sintering area, in air atmosphere).

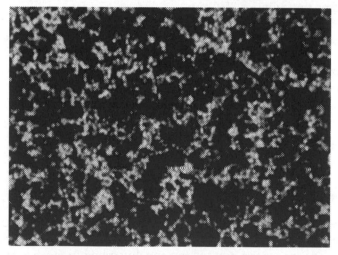

Fig. 5.7   Laser sintering of iron powder with external pressure added in air atmosphere (magnified 100 times, white spots are the pores).

Table 5.1

Micro-hardness	Fe sample in Air	Fe sample in Ar	Pressed Fe sample in air
HV scale	30–12 (HV)	100–80 (HV)	137–96.5 (HV)

Fig. 5.8 Laser sintering of iron powder without external pressure added in Ar atmosphere (magnified 50 times, black spots are the pores).

Fig. 5.9 Laser sintering of iron powder without external pressure added in air (magnified 50 times, white flaky parts are the oxide layer).

Samples

We now consider two samples produced using the system presented. One sample is a small fan-shaped prototype (Fig. 5.10), and the other is a strain gauge prototype (Fig. 5.11). Both prototypes are made from an iron–bronze premix powder in an argon atmosphere.

Fig. 5.10   A small fan prototype.

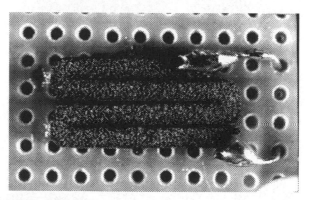

Fig. 5.11   A strain gauge prototype.

5.4 CONCLUSIONS AND FUTURE RESEARCH

During the present research we have observed that low laser power and slow scan are desirable, rather than high power and a fast scan for sintering. Samples that have been sintered with high laser power tend to curl up and melt the powder. It has also been observed that during sintering the samples that form intermetallic compounds are more compact and ductile than pure single phase samples. For our future work we have identified some of the important parameters for sintering. These are: powder/size, laser power, laser spot size, scan speed, layer thickness,

Table 5.1 Comparisons between diode lasers and conventional lasers for material processing

Material process	Intensity, W/cm^2	Interaction time, s (beam dia./travel time, d/v)	Excimer, CO_2, Nd-YAG (10^7–10^{10}), W/cm^2, pulse or CW	Laser diode (LD) (10^3–10^4), W/cm^2, mostly CW
Transforming hardening	10^3–10^4	0.01–1	*	*
Surface melting	10^4–10^6	0.001–1	*	+
Surface alloying	10^4–10^6	0.001–1	*	+
Surface cladding	10^3–10^5	0.1–1	*	+
Particle injection	10^3–10^5	0.1–1	*	+
Laser beam physical vapor deposition (LBPVD)	10^3–10^5	1–6000	*	+
Laser beam chemical vapor deposition (LCVB)	10^3–10^5	1–6000	*	+
Enhanced plating	10^5	100	*	+
Shock hardening	10^8–10^{10}	10^{-8}–10^{-6}	*	–

* Yes, + at present depends on alloy system, – does not work

process environment (Ar, N), pressure on powder/binding powder and scan pattern. With a new optical system, designed and developed by Chen, Roychoudhuri and Banas (1994) it has been possible to have an intensity of 10^6 W/cm^2 from laser diode.

Diode lasers have the potential to be used for many different material processes. Table 5.2 illustrates the potential in prototyping 3D parts; cladding is a melting process. Our next research will be the investigation of laser cladding techniques to rapidly coat the substrate with a thick layer of materials to improve its physical properties (oxidation, corrosion, wear resistance). Due to rapid solidification during laser melting, a cooling rate of 10^{-7}°C can be achieved and may form a nonequilibrium alloy. The nonequilibrium alloys have different electrical, physical and optical properties that cannot be achieved by conventional equilibrium alloys. This technique can also be used to repair and rebuild expensive parts.

REFERENCES

Badrinarayan, B., and Barlow, J. W. (1992) Metal parts from selective laser sintering of metal-polymer powders. *Proceedings of Solid Freeform Fabrication Symposium*, Austin, 141–146.

Banas, C. M., and Webb, R. (1982) Macro-materials processing. *Proceedings of the IEEE*, **70**(6).

Bickford, D. J., and Schenk, J. A. (1993) The rapid prototyping industry in 1993: infant industry structure and dynamics as viewed by industry participants. *Proceedings of the Forth International Conference on Rapid Prototyping*, Dayton, Ohio.

Bohn, J. H., and Wozny, M. J. (1992) Automatic CAD-model repair: Shell-closure. *Solid Freeform Fabrication Proceedings*, TX, 309–330.

Carter, W. T., Jr., and Jones, M. G. (1993) Direct laser sintering of metals. *Proceedings of the Solid Freeform Fabrication Symposium*, Austin, Texas, 135–142.

Chen, W., Roychoudhuri, C. S., and Banas, C. M. (1994) Design approaches for laser-diode material-processing systems using fibers and micro-optics. *Optical Engineering*, **33**(11), 3662–3669.

Crawford, R. H. (1993) Computer aspects of solid freeform fabrication: geometry, process control, and design. *Proceedings of 1993 Solid Freeform Fabrication Symposium*, Austin, Texas, 102–112.

Dong, J. (1996) The issues for computer modeling and interfaces to solid freeform fabrications. *Proceedings of ASME–MED Winter Conference*, Vol. 4, 88–85.

Dong, J., Manzur, T., and Roychoudhuri, C. (1996) Fiber coupled high power laser diodes for solid freeform fabrication directly from metal powder. *Proceedings of 1996 ASME–MED Winter Conference*, Atlanta, Vol.4, 47–53.

DTM Corporation (1992) The selective laser sintering process: third generation desktop manufacturing. *Proceedings of the Third International Conference on Rapid Prototyping*, Dayton, Ohio.

Jacobs, P. F. (1992) Rapid Prototyping and Manufacturing, SME.

Johanson, R. Kikuchi, N., Papalambros, P., Prinz, F., and Weiss, L. (1993) Homogenization design and layered manufacturing of lower control arm in project MAXWELL. *Proceedings of 1993 Solid Freeform Fabrication Symposium*.

Nakagawa, Takeo (1994) Present status of rapid prototyping in Japan. *CIRP Meeting*, Paris.

Roychoudhuri, C., and Chen, Weiqun (1993) New applications of high power laser diodes. *MOC/GRIN'93*, Kawasaki, Japan.

Subramanian, P. K., Zong, G., Vail, N. K., Barlow, J. W., and Marcus, H. L. (1993) Selective laser sintering of Al_2O_3. *Proceedings of Solid Freeform Fabrication Symposium*, Austin, Texas, 350–359.

Uziel, Y. (1992) Stereolithography – a view of the future. *Proceedings of the Third International Conference on Rapid Prototyping*, Dayton, Ohio.

CHAPTER SIX

Beyond product design data: data standards for manufacturing resources

Kevin K. Jurrens, Mary Elizabeth A. Algeo and James E. Fowler

6.1 INTRODUCTION: MANUFACTURING DATA STANDARDS

Standards play an important part in an industry's success or failure in the marketplace. They facilitate common markets, influence marketing patterns in global trade and augment the operations of distributed enterprises.

Standards are formed in a variety of ways [6].

- A particular mechanism that dominates a market may lead to a *de facto* standardization of that mechanism's characteristics.
- A group of vendors may mutually agree to supply a common interface (physical or otherwise) so that their products interoperate, thus creating a standard within their combined customer base.
- An influential user or group of users may require particular features and interfaces in the products of their suppliers or services of their contractors.
- A committee of technical experts may build consensus under the auspices of a public standards-making authority.

The last of these methods results in a formal standard; the results of the others are termed *de facto*, industry, or informal standards. Examples of industry standards include the disk operating system (DOS) for personal computers, military specifications (MIL-SPECs), the DXF data exchange format to share product design data between computer-aided design software systems, or the Caterpillar V-Flange tool holder for cutting tools. Industry standards often serve as a catalyst for developing formal standards.

Formal standards act as a mechanism to insure a degree of compatibility, performance, quality and safety between seller and buyer. With the exception of government-mandated safety standards and standards dealing with direct compatibility (e.g. electrical receptacles, screw threads, gears), compliance with most engineering standards is voluntary. The subjects of such standards are wide and varied, as are the standards-producing organizations. Standards-producing organizations (and their resulting standards) can be either national or international in nature.

National standards are those sanctioned by a country's public standards authority. The American National Standards Institute (ANSI) is the authority that insures the integrity of formal standards developed in the United States. The Deutsches Institut für Normung (DIN) in Germany, the Association Francaise de Normalisation (AFNOR) in France, the British Standards Institution (BSI) in the United Kingdom, and the Japanese Industrial Standards Committee (JISC) in Japan are examples of national standards authorities. The standards development process varies from country to country. In the United States, ANSI does not write standards, but grants national standards-writing accreditation to an organization (i.e. a secretariat) that agrees to follow the ANSI-prescribed *due process*. This due process requires the accredited secretariat to promote the public interest and to insure that the process is voluntary, open and free of self-serving interests. The American Society of Mechanical Engineers (ASME), the Cemented Carbide Producers Association (CCPA), the Electronic Industries Association (EIA), and the United States Product Data Association (US PRO) are a few of the organizations that produce ANSI standards in the domain of manufacturing and/or manufacturing data. Each of these secretariats oversees the formation and administration of committees which address specific standardization subjects. The secretariat seeks a balanced committee membership that represents the interests of all those who have a stake in the standards subject. For manufacturing data standards, representation is sought from manufacturing equipment/hardware vendors, manufacturing software vendors, manufacturing facilities, and research organizations (industry, government and academia). In ANSI committees, each representative member has one vote.

The International Organization for Standardization (ISO) and the International Electrotechnical Commission (IEC) are public authorities that generate international standards. Volunteer national standards authorities (e.g. ANSI, DIN, AFNOR, JISC) which are organizational members of ISO/IEC administratively support individual international standards activities. Technical experts to represent their respective nation's technologies, approaches, practices, and interests staff ISO and IEC standards committees. However, when an issue is brought to a vote in an ISO or IEC committee, each member nation has only one vote. Therefore, it is important for a member nation to come to a consensus within its own constituency. In the United States this task often is accomplished through a Technical Advisory Group (TAG) sanctioned by ANSI and organized by an accredited national standards-writing organization.

Although the process of setting standards varies by organization (and minor differences are possible even between committees within a given organization), several general characteristics can be identified to describe the standards development process. In general, the creation of a formal standard consists of the following steps [6] [27].

- recognizing the need for a standard
- gaining corporate and/or national support for standards investment
- applying existing technology or developing new technology, as appropriate
- developing and documenting a consensus solution
- testing and evaluating the standard
- implementing the standard in industrial environments
- periodically revising or reaffirming the standard to meet changing needs
- retiring or replacing the standard when it reaches obsolescence.

Standards for representing, using and exchanging manufacturing data, such as standards for other aspects of manufacturing, play a significant role in defining and enhancing the capabilities of the manufacturing industry. Several current standards define formal representations for different types of manufacturing data. Examples of some commonly-known manufacturing data standards include the following.

APT (Automatically Programmed Tools, ANSI X3.37 [4]) is a high-level computer language used to generate control instructions for numerically controlled (NC) machine tools and other NC devices. An APT processor reads a part program composed of APT statements and generates a generalized solution in terms of a series of cutter location points that define the cutter path.

RS-274-D (NC file format, ANSI/EIA RS-274-D [2]) specifies an alphanumeric NC file format commonly referred to as M and G codes. This format is used to provide low-level instructions to the controller of an NC machine tool. This format is created by post-processing a cutter location file generated by the NC programming system or APT processor to a form corresponding to the characteristics of the particular machine tool/controller combination.

The BCL (Binary Cutter Location, ANSI/EIA RS-494-B [3]) standard specifies a binary format for data used as input to an NC machine tool. A BCL processor reads a cutter location file generated by the NC programming system or APT processor and converts the data to a form that can be directly used by the machine tool. For NC data formatted using the BCL standard, a post-processing capability is typically embedded in the machine control unit.

DMIS (Dimensional Measurement Interface Standard, ANSI/CAM-I 101 [1]) provides a standard for bidirectional communication of inspection data between computer systems and inspection equipment. Though human-interpretable, DMIS was designed for communication between automated equipment.

STEP (Standard for the Exchange of Product Model Data, ISO 10303 [8]) consists of a series of standards that define an electronic representation for the exchange, sharing and archival of product information. STEP is designed to address data requirements for the complete product life-cycle, from preliminary design through manufacturing, deployment and eventual retirement.

The MMS (Manufacturing Message Specification, IEC 9506 [11]) standards define a network communication protocol to interact with manufacturing equipment. Example capabilities include transferring files, issuing remote commands, and receiving appropriate responses.

With this standards perspective in mind, the remainder of this chapter describes the industrial need for a standardized representation of a specific type of manufacturing data, which we call manufacturing resource data. The chapter presents a case study and proposed solution for a subset of manufacturing resource data as developed by members of the Rapid Response Manufacturing (RRM) Intramural Project at the National Institute of Standards and Technology (NIST), and discusses current prospects for standardization of this type of data. Specific topics include the development process and sources of detailed requirements for an information model to document the proposed data representation, the structure and content of the information model, project interactions with industry and standards groups, plans for implementation and validation of the information model, and the relationship of manufacturing resource data to other product and process data within an integrated environment. Lastly, an overview of other related or similar efforts worldwide and potential future opportunities, are also discussed to illustrate the level of interest and activity in this technical area and to give a sense of magnitude to the proposed effort.

6.2 MANUFACTURING RESOURCE DATA

Manufacturing resources are defined as the equipment, tools, supplies, and facilities which enable industry to perform manufacturing processes. Examples of manufacturing resources include machine tools, cutting tools, inserts, tool holders, tool adaptors, tool assemblies, collets, fixtures, inspection systems and equipment, material handling equipment, welding equipment, coolants, factory layouts, stock materials, etc. Various characteristics about these manufacturing resources must be known and evaluated to make manufacturing decisions. These characteristics are called manufacturing resource (MR) data. (Though the terms *data* and *information* have different meanings in some contexts, the subtlety of their definitions is not captured in this document and the terms are used interchangeably.) Examples of MR data include the cutting diameter of an endmill, the number of axes of a milling machine, the feed direction of a cutting tool insert, the cutting edge material of a cutting tool, and the series designation of a collect. For all practical purposes, the list of possible MR data examples is seemingly endless.

6.2.1 Industry need

In the domain of mechanical design and manufacturing, information sharing between computer-aided design, computer-aided manufacturing and computer-aided engineering (CAD/CAM/CAE) applications is typically synonymous with product design data exchange. The exchange or sharing of product design data typically occurs using existing standards, including the STEP standard, the Initial Graphics Exchange Specification (IGES) standard [5], or the DXF *de facto* standard. Other information elements which are relevant to and necessary for the functions performed by CAD/CAM/CAE applications are frequently overlooked. For example, information that describes machine tool capabilities is necessary for manufacturing planning and simulation purposes; information that describes specific process characteristics is useful for analysis and cost estimation purposes. It is clear that CAD/CAM/CAE applications require such manufacturing resource information and employ computer-interpretable representations of the data. It is equally clear that each CAD/CAM/CAE vendor maintains such MR information in subtly different ways. Moreover, the originating sources of such information (e.g. machine and cutting tool manufacturers) frequently do not provide equivalent characterizations of their products. With this current environment, CAD/CAM/CAE system integration or the sharing of MR data between systems or engineering functions is not possible without significant levels of effort and/or loss of information.

To illustrate this point, Fig. 6.1 and 6.2 provide screen views from two sample manufacturing process planning applications that require the use of MR data. For each example, the attributes of a common twist drill as stored and used by that application are shown. Comparison of these two figures demonstrates that the set of MR data attributes maintained by the system in Fig. 6.1 are somewhat different to the attributes maintained by the system in Fig. 6.2. Though these two application systems are similar in functionality (i.e. both are used for manufacturing process planning), the sharing of MR data between the two systems is not possible without

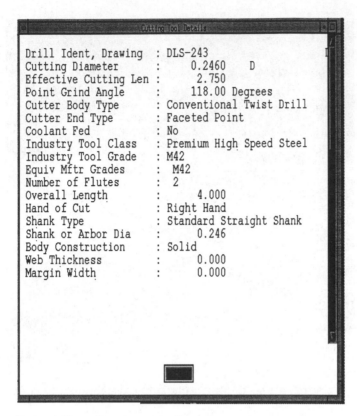

Fig. 6.1 Twist drill attributes, system 1.

information loss. Similar situations can be shown for numerous other CAD/CAM/CAE applications.

Manufacturing resource data is typically represented multiple times in multiple formats for different engineering applications within a given facility. This situation results in redundant data stores of the same or similar information, much duplicate work for maintenance of the data, lack of quality control to ensure that the most recent and accurate resource data are used by all personnel, and slower system implementations for new applications that require the use of MR data.

The representation of manufacturing resource data has a substantial impact on manufacturing operations. Several engineering and manufacturing functions (with or without the benefit of automated application software systems) require access to accurate, current, and complete sources of MR data. Primary examples of these functions include: manufacturing process planning, NC code generation, tool management, tool and fixture design, material requirements planning (MRP),

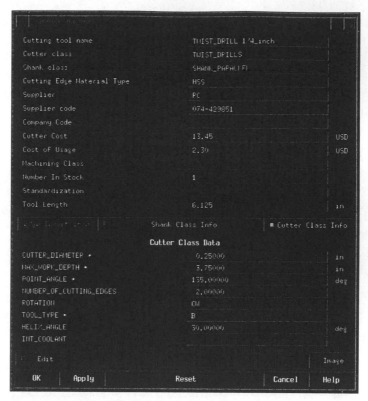

Fig. 6.2 Twist drill attributes, system 2.

manufacturing simulation, manufacturing cost estimating, inspection planning, facility planning and production scheduling.

This impact carries through to the manufacturing personnel who perform these functions, the commercial software vendors who develop automated tools for these functions, and the in-house software implementors who design company-specific software systems.

Several efforts within industry and academia have developed information constructs (e.g. databases and information models) that describe MR data. These efforts, however, have typically resulted in company-specific data structures and system implementations. The implementations are frequently applicable to only a single CAD/CAM/CAE application within the organization, with much duplicated effort required to implement systems for additional application areas. These efforts appear to have been influenced heavily by specific sets of vendor software in use within that organization.

6.2.2 Benefits of a standardized MR data structure

A standardized electronic representation of manufacturing resource data would define the categories of MR data, the attributes of each MR data type and the relationships between MR data types (i.e. the information structure). The values of the MR data attributes generally would not be included in this representation, except where required for defining structural elements of the representation or when all known (possible) values of the attribute can be enumerated. Typically, a standardized electronic representation is documented in the form of an information model (or data model). The information model documents the data requirements in a structured and unambiguous form for consistent interpretation by users and implementors. Although several information modeling languages and methods exist, further elaboration on information modeling techniques is beyond the scope of this chapter.

The motivation for developing complete, validated, and publicly available specifications for the representation and exchange of MR data can be viewed from multiple perspectives. In general, such standards would ensure a degree of compatibility, performance and quality among manufacturing facilities, manufacturing resource providers and manufacturing software suppliers. The development of validated MR information models would serve as a catalyst to standardize these models by providing proven results and a working strawman to appropriate standards organizations.

From the manufacturing facility perspective, a standardized MR data structure will provide the ability to share MR information among engineering functions and allow more seamless integration of manufacturing software applications. Shared MR data yields a cost savings by reducing the amount of effort required for the maintenance of redundant stores of tooling data. In addition, shared MR data increases quality control by ensuring that the most recent and accurate resource data is used by all personnel. New applications that require the use of MR data can be implemented more efficiently and quickly. A consistent characterization of manufacturing resources also enables more direct comparison of vendor products. Finally, as a customer, a manufacturer may ultimately obtain resource data in a common electronic format directly from tooling vendors, thereby eliminating the tedious and time-consuming manual entry of such data.

For manufacturing resource providers (e.g. machine tool vendors and tooling vendors), a standardized MR data structure will provide the ability and incentive to disseminate specifications of their products via mechanisms consistent with state-of-the-art computing and networking technologies, such as electronic catalogs, integrated databases, CD-ROM disks or on-line services. Manufacturing resource data standards will facilitate the establishment of electronic MR data books and catalogs for use throughout the manufacturing industry. A standard MR data format enables vendor product information to be directly imported into customer databases or applications. In addition, the possibility for enhanced collaboration between MR providers and manufacturing software vendors seems likely. New opportunities for competitive advantage could exist (initially anyway) for those MR vendors providing electronic product representations using the standardized form. (Of course, customers will still consider aspects such as performance, quality, applicability to their needs, and vendor support in making their selections. Other existing competitive advantages between suppliers, such as functionality, special materials, geometry, or proprietary mechanisms, will still exist.) As the practice becomes more widespread, however, this advantage will be gradually reduced.

The availability of a standard MR data structure will allow CAD/CAM/CAE software developers and research organizations to concentrate on application-specific problems or the goals of the research, rather than acquisition, representation and maintenance of MR data. This leads to more efficient use of research funding and less duplication of work to supply MR data to multiple applications. Software developers will have increased ability to integrate with other vendor software and may recognize extensions or additional capabilities to include in their applications due to the availability of MR data structures and potential liaisons with MR providers.

6.3 CASE STUDY: DEVELOPMENT OF A PROPOSED SOLUTION

The following case study provides details of a proposed solution for the standardization of a subset of MR data as developed by members of the Rapid Response Manufacturing (RRM) Intramural Project at the National Institute of Standards and Technology (NIST). This case study is provided as an example to illustrate the development process and content of the resulting data structure. It must be emphasized that the MR data representation presented in this case study is a strawman proposal for standardization and is intended to serve as a starting point for a future standardized MR data representation. This initial proposal will be modified and refined through further review, implementation experience, and the standards process. This proposed structure should not be considered the only possible solution.

6.3.1 Rapid Response Manufacturing (RRM) industry consortium

The rapid response manufacturing program was established in 1992 to 'provide the effort needed to effectively enable engineers to reduce the time required to design and manufacture products in response to rapidly fluctuating market demands by one-half' [21]. The objective of the program is to shorten time-to-market, improve quality-to-cost, and enhance product reliability. The RRM program is a five-year, $50 million, 10-company program that is coordinated through the National Center for Manufacturing Sciences (NCMS). The program is co-sponsored through the Department of Commerce Advanced Technology Program (ATP) administered by NIST and matching funds from industry participants. Activities of the RRM consortium include developing new processes, adding new functionality to CAD/CAM/CAE software applications, and improving system and process integration among various engineering functions and software applications [25].

The RRM industry consortium is composed of both manufacturing organizations and software vendors. Manufacturing organizations within the consortium include the General Motors Corporation (Detroit, MI), the Ford Motor Company (Dearborn, MI), Texas Instruments (Dallas, TX), the United Technologies Corporation (South Windsor, CT) and Lockheed Martin Energy Systems (Oak Ridge, TN). Each manufacturer has chosen a specific part (or part family) to focus their RRM efforts, as shown in Table 6.1 [15] [19]. The following software vendors are also included in the RRM consortium as participating partners: the MacNeal-Schwendler Corporation (Los Angeles, CA), Spatial Technology (Boulder, CO), Teknowledge (Palo Alto, CA), CIMPLEX (San Jose, CA) and Concentra (Cambridge, MA).

The RRM program has initiated more than 70 projects in various technical areas since its inception [22]. These projects have addressed diverse subjects such as manufacturing planning, variant product design, tolerance analysis, geometric and feature representations, knowledge-based engineering systems, rapid tooling and NC toolpath generation.

One RRM industry consortium effort of particular interest to this discussion is the integrated product and process model (IPPM) concept. A stated goal of the RRM program is to create an electronic representation

Table 6.1 NCMS RRM industry consortium

Manufacturer	Division	Part family
General Motors	Saginaw	Steering shaft component
Ford Motors	Powertrain	Crankshaft
Texas Instruments	Defense Systems and Electronics Group	Complex housing case
United Technologies	Pratt & Whitney	Fan casing
Lockheed Martin	Energy Systems	Turned parts

to capture both product and process information in a unified manner (i.e. the IPPM). The IPPM would be used to support the software integration of a selected set of engineering applications from product design through production. Product data within the IPPM would include aspects such as geometry, topology, tolerances, material, form features, product config- uration, etc.; primarily the information contained in the ISO 10303 STEP standards. Process data would extend the IPPM to include several additional types of manufacturing data, such as process characteristics, process control data, quality data, NC machine instructions, scheduling information, manufacturing plans and manufacturing resource character- istics. Due to the magnitude of the overall IPPM scope, the model is being developed and extended in stages [19].

6.3.2 NIST Rapid Response Manufacturing (RRM) Intramural Project

The NIST RRM project [23] is sponsored as an intramural project through the NIST ATP office to support the objectives of the NCMS RRM consortium. The NIST RRM intramural project is managed and executed through the Manufacturing Systems Integration Division (MSID). The RRM intramural project supports RRM program efforts by leveraging NIST skills and technologies to ensure the advancement of RRM capabilities, collaborating to develop and adopt key technologies, and providing a standardization focus to the results of the RRM consortium. Specific activities of the RRM intramural project are determined by research and technology needs identified by consortium member companies.

One of the primary activities of the NIST RRM intramural effort has been the development of a representation for manufacturing resource data. This work was initiated in the context that MR data is one aspect of an overall IPPM that had not yet been addressed in a comprehensive manner. Results of this work are expected to integrate with other product and process data elements within the IPPM developed by NCMS RRM program participants.

The primary goal for this RRM intramural activity is to assess the feasibility of a standardized MR data structure by demonstrating the existence of a common set of manufacturing resource data to support the functions performed by a variety of CAD/CAM/CAE software applica- tions. Specific objectives of this effort include:

- to develop information models that specify data requirements for a subset of manufacturing resource data
- to evaluate the completeness, validity and usability of the information models by implementing in a shared database environment to support multiple CAD/CAM/CAE applications

- to translate the results of the manufacturing resource modeling into resources for related research and development efforts (e.g. NCMS RRM industry consortium, other industry or academic efforts, internal NIST efforts)
- to use the project results to provide a catalyst for a standardized manufacturing resource data structure.

The development methodology for this project consisted of the following steps.

1. Determine the MR data scope to be addressed by the project.
2. Gather and analyze MR information requirements from multiple sources and perspectives.
3. Develop and document the proposed MR data representation in the form of a detailed requirements specification.
4. Disseminate the requirements specification for widespread industry review by manufacturers, MR vendors, manufacturing software developers and members of standards organizations.
5. Develop an information model based upon the content of the MR requirements specification using the EXPRESS [9] modeling language.
6. Implement the EXPRESS model into a commercial database system (using automated software tools) and populate the database with industry MR data.
7. Evaluate the proposed data representation through integration of commercial manufacturing software applications with the MR database (i.e. the MR Data Test Environment).
8. Update the MR requirements based on the results of the industry review and validation efforts.
9. Interact with technical committees from national and international standards organizations to assess interest and applicability for standardization of MR data.
10. Submit project results to appropriate standards organizations as a strawman proposal for standardization.

6.3.3 RRM intramural MR data scope

Several factors influenced the determination of a scope for the MR data analysis. An essential factor was the manufacturing methods of interest to the NCMS RRM consortium members. In their joint efforts, consortium members were largely focused on improving the design to production cycle time for parts produced by milling, drilling and/or turning. The workpiece materials for the machined components within the consortium part families included aluminum, carbon steel, nickel-based steel and titanium. Based on these constraints, the following manufacturing resource types were deemed to be 'in-scope'. (Information about fixturing components and mechanisms (e.g. vises, modular tooling, custom jigs) was also deemed important. However, project time constraints necessitated that fixture information be left for future work.)

- Machine tools
    N-axis milling machines, processes – mill, drill, bore, ream, tap, saw
    N-axis turning machines, processes – turn, bore, thread, groove, face, profile, cut-off
    Combination milling/turning machines.
- Cutting tools: mills, drills, taps, countersinks, counterbores, reamers, boring tools, lathe tools including solid cutting tools, insertable tools and brazed tools.
- Inserts.
- Cutter assembly components: clamps, screws, shims, chipbreakers, interchangeable pilots, etc.
- Tool holders: adaptors, arbors, collets, bushings and chucks.
- Tooling assemblies.
- Tooling assembly components: collet nuts, spacers, retention knobs, arbor retaining bolts, etc.
- Cutting tool materials.

The simplifying assumption was made that the starting state of the work-in-process would be 'ready-stock,' meaning that no blanking or stock preparation processes would be considered. Based on these constraints and assumptions, the following manufacturing operations/technology areas were deemed to be 'out-of-scope': grinding, casting, electrical discharge machining, extrusion milling, band sawing, sheet metal parts and processes, laser cutting, water jet cutting, welding, broaching, planing, and modular/proprietary 'quick-change' tooling. (Specifically, the vendor-specific interface mechanisms used to couple 'quick-change' tool components together are not within the project scope.)

Another factor that affects the scope of the MR data analysis was the proportion of the product lifecycle or production process considered (e.g. material resource planning, factory scheduling, off-line inspection, etc.). The project decided that the manufacturing engineering activities within the production cycle to be considered would be limited to manufacturing cost estimating, manufacturing process planning, and NC program generation. This decision was based on several factors, notably the detailed use of manufacturing resource information that these processes require. Among the consequences of this decision are that information relating to the status of equipment for scheduling purposes is not covered, and administrative information which might be used for accounting, inventory, or maintenance purposes is also not covered.

Finally, the scope of the information addressed by the RRM intramural project is not intended to be specific to United States (US) equipment, practices, or standards. A conscious effort has been made to ensure that a broad spectrum of international resource information has been taken into account. At the same time, the project cannot guarantee that every applicable US or European or other manufacturing resource style, type, usage, or standard has been examined. The project endeavors to seek widespread review of the proposed data structure with the hope that further enhancements to the information covered will result.

6.3.4. RRM intramural MR requirements development

Multiple representations of existing MR data were analyzed to review the current state of the-practice and to form a baseline from which to begin. MR data representations in resource vendor product specifications, manufacturing handbooks, technical journals, CAD/CAM/CAE application software documentation, industry-developed databases, as well as national and international standard documents were examined in detail during the course of this effort. The project also sought, and continues to seek, expertise from practitioners in the field of manufacturing equipment production and use. In general, the project acquired a large amount of raw information about manufacturing resources in the domain of interest. A great deal of effort was required to analyze the raw information such that commonalities could be identified and accurately compared. The sources of MR information and issues relevant to the particular analysis efforts are discussed in the following sections.

Manufacturing software applications

Several CAD/CAM/CAE application software systems were obtained and examined by the project for the identification of MR requirements, as well as for future testing of the MR information model. The internal data structures of these software applications were analyzed (either through system documentation or actual execution of the software) to determine which MR data elements are captured, used, and/or created by the software to perform its stated function. These software systems are summarized below.

Pro/Engineer (Parametric Technologies Corporation) is a computer-aided design system. One of the many add-on modules available for the system is Pro/Manufacture which provides capabilities for NC toolpath generation-based on part features. MR data to describe machine and cutting tools are defined by the user in Pro/Manufacture to assist in the generation of the toolpaths.

MetCAPP (Institute of Advanced Manufacturing Sciences, Inc.) is a manufacturing process planning system. The user identifies work-stations/machines, set-ups, routing sequences and generic part feature data. The system selects cutting tools, calculates feeds and speeds, processing times, and produces process plan and operation sheets. The system's calculations are dependent on detailed descriptions of machine and cutting tools.

ICEM PART (Tecnomatix Machining Automation) is a manufacturing process planning system and NC toolpath generation system. The system accepts an input file describing part geometry. This part geometry is analyzed to automatically identify manufacturing features. The user provides detailed descriptions of stock material, machine tools, holders, adaptors, cutting tools and fixtures which the system uses to automatically generate set-ups, operation sequences, toolpaths, feeds, and speeds.

Manufacturing Analyst (CIMPLEX Corporation) is a manufacturing process planning and NC toolpath generation system. The system accepts an input file to describe part geometry and features. The system has the ability to recognize certain geometries and features and can generate NC toolpaths automatically for these aspects. Mechanisms are also provided for creation of user-defined toolpaths. The software requires descriptions of machine and cutting tools to determine sequences, feeds, speeds, and toolpaths.

SmartCAM (Structural Dynamics Research Corporation) is an NC programming system. The system accepts an input file to describe part geometry. The user manually defines toolpaths based on the geometry of the part and stock. Feeds and speeds are selected by the user based on knowledge of part material, cutting tool details, finish desired and other considerations. Descriptions of cutting tool parameters are needed by the system to define the toolpaths.

The project examined the functions of these software applications, analyzed the system and user documentation, experimented with the system operations, and ultimately compared the resource needs of each software application with respect to each other. All the applications required cutting tool descriptions. The specific cutting tool details required by each system varied. Not all the applications required machine tool descriptions to perform their functions. For those that did, again the specific machine tool details required by each system varied. The majority of the software applications did not require any substantive information about tool holders, adaptors, or tooling accessories to perform their current functions. It is expected that future software capabilities could be enhanced through additional access to and use of these types of manufacturing resource data.

In the comparison of the different software applications, the project uncovered several examples of different nomenclature for what was determined to be the same information. The project also found it difficult to determine whether all resource information identified was actually being used by the systems. Thus it was unclear whether certain resource information was being required by the software for 'completeness', as compared with being necessary for calculation of feed rate, for example.

In the final analysis, the project strove to include in the requirements specification only those resource details for which there was a clear understanding of why those details were necessary.

Manufacturing resource vendor specifications

The project analyzed a considerable number of catalogs, brochures, and technical specifications from a variety of MR vendors and distributors. A small sample of the organizations includes: Cleveland Twist Drill, GTE Valenite Corporation, MSC Industrial Supply Company, Sandvik Coromant Company, Teledyne Cutting Tools, Giddings & Lewis, Inc., Boston Digital Corporation, Command Corporation, Kennametal, Inc., OSG Tap and Die, Inc., SGS Tool Company, The Weldon Tool Company, Cincinnati Milacron Corporation, and Hurco Manufacturing Company.

The project examined the literature available from these sources to analyze how resources were classified, what specifications were presented for each of the resources in scope, which information about a given resource was needed to assemble it with other resources, and the manufacturer's recommendations for the selection and application of resources. All these information elements were accumulated and compared with those already gleaned from the CAD/CAM/CAE software applications. In many cases the CAD/CAM/CAE application requirements were more detailed than those provided in the vendor literature. Again, nomenclature differences among vendors and between vendor literature and the CAD/CAM/CAE documentation often proved confusing. With respect to cutting tools in particular, the characterization of solid tools versus the combined characterization of inserts and holders yielded much lively debate.

Discussions with some of the MR vendors allowed project staff to better understand issues relevant to resource specification. MR vendors were interested (and sometimes surprised) to learn of the detailed resource information called for by the manufacturing software applications.

Industry tooling databases

The project established liaisons with several industrial manufacturing facilities that use the manufacturing resources of interest on a daily basis. Some of these users have created their own internal models of MR information and maintain this data in commercial database systems. RRM Intramural staff had the opportunity to analyze documentation of these resource databases. In particular, MR information from Texas Instruments Corporation and Allied-Signal Aerospace was quite detailed and beneficial. Once again, the information gleaned from these users was compared with the information maintained by the manufacturing software applications and with the information specified by the MR vendors. The comparison revealed that the industry databases often contained a superset of the MR information that had previously been identified. The need for this additional information was typically attributed to the company's multiple uses of the database, e.g. for tool inventory management, for tool selection, for administrative record keeping, for maintenance logging, and so on. In addition, some industry users have developed custom applications that have functionality and resource data needs beyond that usually provided by commercial systems. The project determined that the in-scope data maintained in these databases were consistent with those which had previously been identified. The industry user databases also helped with the task of characterizing cutting tool inserts and turning tools.

Manufacturing standards

Additional sources of MR data requirements were manufacturing-related standards from several national and international standardization organizations. Although several organizations produce standards relating to manufacturing resources, the project specifically reviewed documents from ISO and ANSI. ISO Technical Committee (TC) 29 and TC39 develop and publish standards on small tools and machine tools, respectively. Some of the primary developers of US national standards for machine tools and tooling components are the American Society of Mechanical Engineers (ASME), the United States Cutting Tool Institute (US-CTI), the Electronic Industries Association (EIA), and the Cemented Carbide Producers Association (CCPA). These organizations typically develop technical specifications for sponsorship as a national standard through ANSI. Committees within these organizations focus on the following areas:

- ASME Committee B5: components, elements and performance of machine tools
- ASME Committee B94: cutting tools, holders, drivers and bushings
- CCPA Committee B212: carbide cutting tools, tool holders and test methods
- US-CTI: cutting tools and tool holders
- EIA Committee IE-31: numerical control systems and equipment

The standards review began with a search of the ISO and ANSI catalogs. Due to the large number of relevant standards, the review was accomplished with a subscription to an electronically based document service. This subscription consisted of a collection of CD-ROMs that contained the full text of ISO, ANSI and ASME standards related to the fields of manufacturing and mechanical engineering. Using the classification structure and search methods provided with the CD-ROMs, the project located, viewed, and analyzed more than 150 relevant standards.

The content of the standards regarding manufacturing resource attributes varied significantly. Standards that specify general nomenclature and terminology often identified various subtypes of manufacturing resources as well as their attributes. However, this type of standard did not exist for all types of manufacturing resources. Therefore, attribute information was extracted from tables in standards which list specific values for a given resource and/or from standards that specify performance test attributes and methods. In many cases, the standards specified attributes that may be required for any engineering application (from tool design to tool selection). The project identified those attributes required by the CAD/CAM/CAE applications within the project scope.

Other sources

In addition to the aforementioned sources of resource information, the project obtained useful background, guidance, and insights from the following sources:

- relevant research projects and resource characterizations performed by industry and academic organizations
- discussions with production staff and manufacturing experts from a variety of companies and perspectives
- a variety of published literature and engineering reference books.

The reader is referred to [16] for a comprehensive listing of MR data sources and references.

6.3.5. RRM intramural results and details of proposed solution

After having collected a large body of source material for MR data requirements, the RRM intramural project collated the raw data into a comprehensive specification which formally categorized the various MR elements, defined attributes of interest with respect to each MR element, and described the relationships between MR elements [16]. A formal information modeling method, such as the EXPRESS language, could have been used to describe the collated MR information. However, the project wanted to ensure that the documented MR data requirements could be reviewed and evaluated by the widest possible audience, i.e. manufacturing personnel from a variety of backgrounds, not only those who might be fluent in the EXPRESS language or be experienced with another modeling methodology.

Hence, the MR data requirements were captured in the form of tables and relationships between tables. Each table in the specification corresponded to a major MR data element, e.g. a tool assembly, a twist drill or

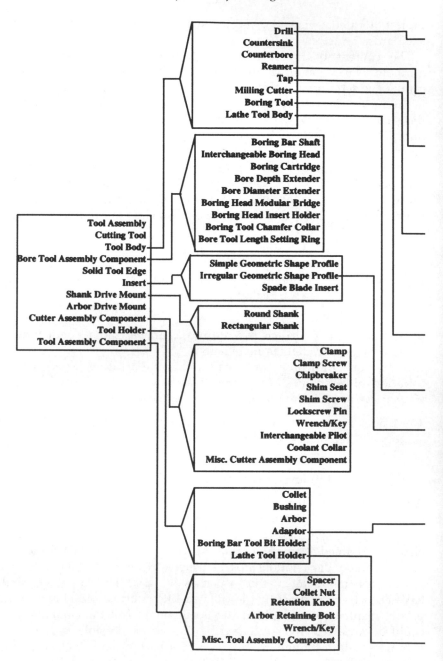

Fig. 6.3 Tooling assembly resource hierarchy.

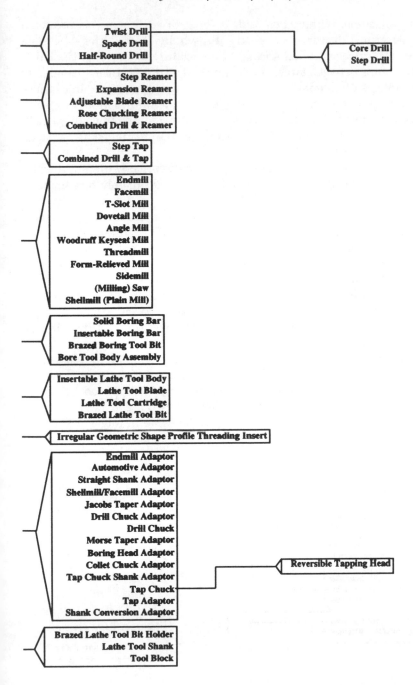

Twist Drill
Spade Drill
Half-Round Drill

Core Drill
Step Drill

Step Reamer
Expansion Reamer
Adjustable Blade Reamer
Rose Chucking Reamer
Combined Drill & Reamer

Step Tap
Combined Drill & Tap

Endmill
Facemill
T-Slot Mill
Dovetail Mill
Angle Mill
Woodruff Keyseat Mill
Threadmill
Form-Relieved Mill
Sidemill
(Milling) Saw
Shellmill (Plain Mill)

Solid Boring Bar
Insertable Boring Bar
Brazed Boring Tool Bit
Bore Tool Body Assembly

Insertable Lathe Tool Body
Lathe Tool Blade
Lathe Tool Cartridge
Brazed Lathe Tool Bit

Irregular Geometric Shape Profile Threading Insert

Endmill Adaptor
Automotive Adaptor
Straight Shank Adaptor
Shellmill/Facemill Adaptor
Jacobs Taper Adaptor
Drill Chuck Adaptor
Drill Chuck
Morse Taper Adaptor
Boring Head Adaptor
Collet Chuck Adaptor
Tap Chuck Shank Adaptor
Tap Chuck
Tap Adaptor
Shank Conversion Adaptor

Reversible Tapping Head

Brazed Lathe Tool Bit Holder
Lathe Tool Shank
Tool Block

a milling machine. Within each table a series of attributes was provided which described features necessary to wholly define the MR data element. The project defined special nomenclature to allow for depiction of relationships between tables, inheritance of attributes from one table to another, ranges of acceptable values for attributes and so on. In essence, these rules provided the bare essentials of an information modeling methodology needed to formally specify the requirements in a self-contained manner without over-burdening the reader with a steep learning curve.

The use of inheritance was fundamental to the organization of the raw resource requirements into a resource hierarchy. Ultimately two separate (but related) hierarchies were developed, one for tool assembly (i.e. cutting tools, tool holders, adaptors, etc.) and the other for machine tools (i.e. milling machines, turning machines and their functional components). The hierarchical organization for tooling assembly and machine tool resource tables is illustrated in Fig 6.3 and 6.4, respectively. In these figures each name corresponds to a table in the data requirements specification; the lines depict 'parent' tables and their descendants. The table furthest to the left in the diagrams have no parent tables, whereas the tables furthest to the right have several ancestors and thus inherit attributes from each of them.

In addition to organizing the resource data into a hierarchy, it was also necessary to determine how data elements related to each other in practice. For example, a tool assembly could be composed of a tool holder, a collet, a cutting tool, and a collet nut (see Fig. 6.5). However, a tool assembly is not always composed of the same number or types of

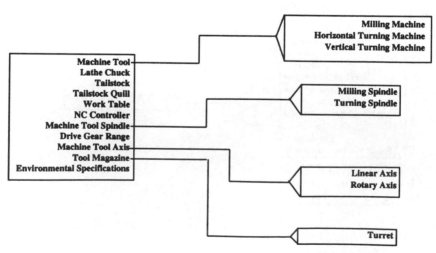

Fig. 6.4 Machine tool resource hierarchy.

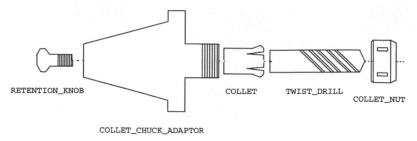

RETENTION_KNOB   COLLET   TWIST_DRILL

COLLET_NUT

COLLET_CHUCK_ADAPTOR

Fig. 6.5 Sample tool assembly.

components. Hence it was necessary to provide a mechanism which allowed for variable compositions of resource data tables. The resource specification provided for such compositions through the use of a special nomenclature that allowed the aggregation of references between tables as well as differentiating between required and optional references between tables.

Such references were used in several tables to establish connections between resources. The approach minimized redundancy in table attributes while at the same time reflecting how resources relate to one another. Figure 6.6 identifies the resource tables comprising a tool assembly and the structure of the references between the tables defined in the resource specification. The arrows in the figure are oriented such that the head of the arrow points at the table being referenced. When a variable number of table references was required, the arrow head is labeled with the word LIST; otherwise the arrow is unlabeled to indicate that a single table reference was specified. Arrows drawn with a broken line indicate that the table reference was OPTIONAL; arrows drawn with a solid gray line indicate that the table reference was CONDITIONAL, i.e. dependent on the value of another attribute or some other information. The notation ONE OF in Fig. 6.6 indicates that a selection of options was required.

The relationships between the tables comprising a TOOL_ASSEMBLY were relatively compact compared with the tables which comprised a MACHINE_TOOL. The complexity of relationships between tables characterizing aspects of machine tools undoubtedly stemmed from the fact that most contemporary machining centers can be configured to perform a wide variety of operations.

Figure 6.7 illustrates the resource tables and the structure of the references between tables for four areas of the machine tool resource hierarchy. The upper left view of the figure depicts the MACHINE_TOOL table with its references to NC_CONTROLLER and ENVIRONMENTAL_SPECIFICA-TIONS tables. As MACHINE_TOOL is the parent table for all specific types of machines, each of its subtypes will also include those two references as

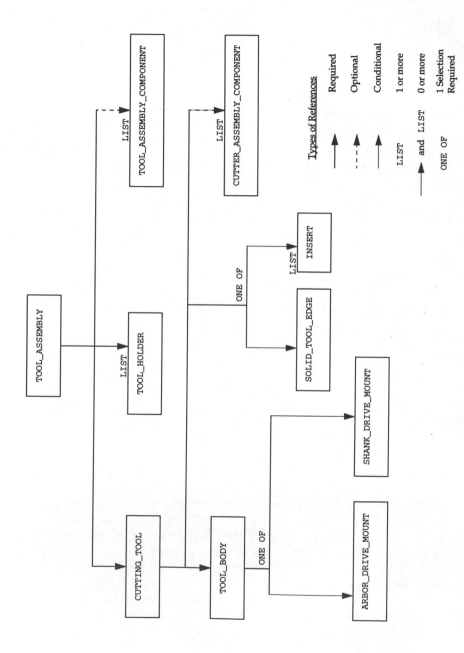

Fig. 6.6 Tool assembly and cutter composition.

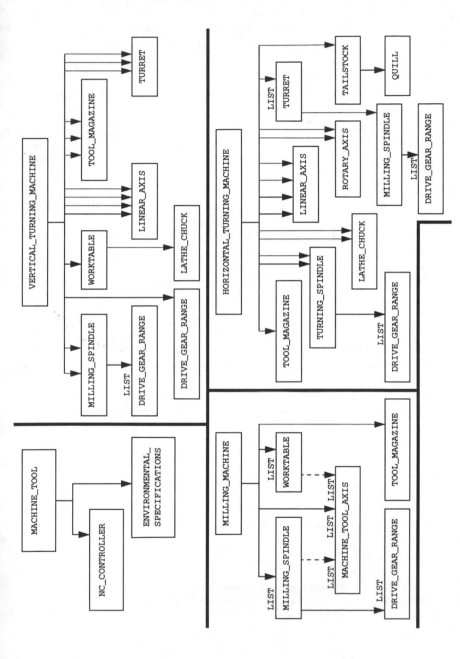

Fig. 6.7 Machine tool table compositions.

well. The three other views in the figure depict table relationships unique to each subtype of MACHINE_TOOL. Due to the flexibility provided in the resource tables for characterizing VERTICAL_TURNING_MACHINE, HORIZONTAL_TURNING_MACHINE, and MILLING_MACHINE, the illustrations in Fig. 6.7 represent most, but not all, of the possible table references involved in their characterizations.

On completion of the MR requirements specification, the RRM intramural project turned to the development of an information model to more formally document the proposed MR representation. This information model was developed using the EXPRESS data description language. The reasons for creating this information model were two-fold. First, standardization of electronic data formats typically requires the use of a formal information modeling language. The RRM project felt that an EXPRESS model of the MR requirements would facilitate acceptance of the proposed structure by standards organizations. Secondly, the project planned to assess the utility of the proposed data structure by implementing an MR database corresponding to the resource tables defined in the requirements specification. Rather than manually creating the structure for the MR database, the project wished to make use of existing NIST and commercial tools for developing the structure of a database automatically from an EXPRESS model.

Although laborious, converting the MR table structure to an EXPRESS model was relatively straightforward because the resource tables were defined using formal rules consistent with the modeling capabilities of EXPRESS. Tables from the requirements specification (generally) became entities in the EXPRESS model. Figure 6.8 illustrates the EXPRESS entity definition for a sample entity, TOOL_ASSEMBLY. The RRM project also created the companion graphical form of the EXPRESS model using the EXPRESS-G language. The graphical EXPRESS-G version of the example TOOL_ASSEMBLY entity is shown in Fig. 6.9.

```
ENTITY tool_assembly;
    tool_assembly_information : OPTIONAL SET [1:?]
        OF tool_assembly_component;
    tool_assembly_number : identifier;
    tool_holding_information : LIST [1:?] OF
        tool_holder;
    maximum_assembly_diameter : size;
    overall_length_of_assembly : length_measure;
    tool_offset_length : length_measure;
    depth_capacity : length_measure;
    magazine_slot_number : INTEGER;
    cutting_tool_information : cutting_tool;
END_ENTITY;
```

Fig. 6.8   Sample EXPRESS definition.

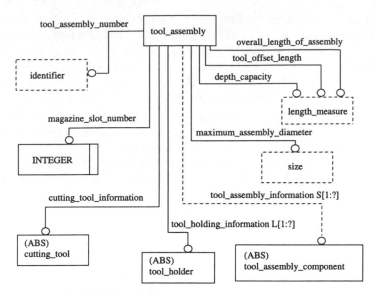

Fig. 6.9   Sample EXPRESS-G definition.

6.3.6 RRM Intramural evaluation of proposed MR representation

With the MR requirements specification fully documented, it was then possible to begin the task of evaluating the accuracy, completeness and usefulness of the information. This effort began with widespread dissemination of the requirements specification document to acknowledged experts in the manufacturing field to seek their assessment. The MR requirements specification was distributed for review to manufacturers, MR vendors, manufacturing software developers, researchers, and members of standards organizations. Results of these external assessments have shown the requirements specification to be comprehensive with regard to the MR types covered.

Manufacturing users, including members of the NCMS RRM consortium and the US Department of Energy-sponsored Technologies Enabling Agile Manufacturing (TEAM) consortium, have indicated that although company-specific data elements are missing, the proposed structure generally contains a thorough and accurate set of requirements. Manufacturing software vendors have indicated that the proposed representation would prove beneficial, especially regarding the tool-holding components and the interface between components of the tooling assembly. Academic researchers, from the University of Michigan and Florida International University in particular, have analyzed the proposed data structure in detail and have incorporated aspects of it into their work. Other reviews are currently ongoing.

With the completion of the MR EXPRESS model, further plans for validation of the data structure through database implementation and application integration could be carried out. The RRM intramural project developed a manufacturing resource data test environment consisting primarily of a commercial object-oriented database system and manufacturing software applications that require use of MR data (see Fig. 6.10). The initial set of manufacturing applications included in this test environment were manufacturing process planning, manufacturing cost estimating, and NC toolpath generation systems.

The MR EXPRESS model was implemented into the database system using automated software tools. These tools automatically generate the schema for the database based on the EXPRESS structure. The database

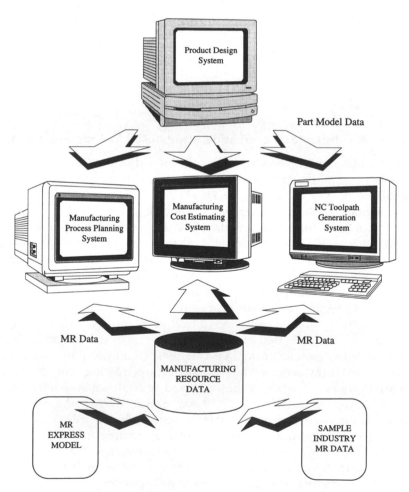

Fig. 6.10 Manufacturing resource data test environment.

was then populated with sample industry and vendor manufacturing resource data. This exercise in itself provides feedback on the completeness and usability of the model. Mapping of the existing industry and vendor data to the proposed structure illustrates potential problem areas and provides suggestions for improvement.

Integration of commercial manufacturing software applications with the MR database is an area of current work within the project. It is expected that these efforts will demonstrate multiple applications functioning from the same set of resource data. This integration will require modifications within the software systems to obtain MR data from the RRM MR database, rather than from other (probably internal) sources. This demonstration would prove that, from the particular application perspectives, the MR data structure was sufficient to allow the sharing of MR data and execution of the software application. The MR data test environment and demonstrated results also provide key elements in the 'marketing' and acceptance of the proposed solution and of the concept of a standardized MR data representation in general.

The final method of validation pursued by the RRM intramural project is consensus-building through standards organizations. The project has contacted members of various technical committees from national and international standards organizations to assess the committee's interest and applicability for standardization of MR data. In addition, the RRM requirements specification was distributed to representatives of standards groups with a potential interest in standardization of this type of data. Project results have been offered as a strawman proposal for standardization, with much interest generated at both a national and international level. Further discussion on standardization of MR data, including general issues and specific efforts of various committees, is presented in the following section. NIST RRM intramural project results regarding standardization are incorporated in this discussion.

The resultant of the RRM evaluation efforts will lead to modifications to the proposed MR requirements and electronic representation. These updates will be as a result of the industry review, implementation experience, and the standards consensus process. Ultimately, all feedback will be forwarded into the appropriate standardization efforts.

6.4 MR DATA STANDARDIZATION EFFORTS

Given the global nature of manufacturing resources and related markets, as well as the presence of US companies in those markets, the need for standardization of MR data exists in both the international (ISO) and US national (ANSI) arenas. As the topic of MR data representation and exchange spans multiple technical domains, standardization efforts are being discussed and pursued in several standards committees. Of particular interest are those committees that focus on specific types of manufacturing resources (i.e. the hardware) and those that address the application of computer technology to manufacturing (i.e. the information).

From the hardware side, the activities of ISO Technical Committees TC29 (small tools) and TC39 (machine tools) and ANSI committees ANSI/ASME B94 (cutting tools), ANSI/CCPA B212 (carbide tooling) and ANSI/ASME B5 (machine tools) will provide requirements for standardization of MR data. From the manufacturing information perspective, ISO TC184 Subcommittee (SC) 4 (industrial data and manufacturing languages) and the ANSI/USPRO IGES/PDES Organization (IPO) (product data representation and exchange) are expected to play a role in the standardization of MR data. PDES (Product Data Exchange using STEP) is the US project in support of the international ISO TC184/SC4 standards. (ISO TC184/SC4 standards include STEP (ISO 10303), Parts

Table 6.2 Manufacturing resource and related standards

General subject	*International standards committees*	*US standards committees*
Manufacturing resources	ISO TC29 small tools	ANSI/ASME B94 cutting tools, holders, drivers and bushings
		ANSI/CCPA B212 cemented carbide, oxide, and diamond cutting tools for turning and/or boring; tool holders; test methods
	ISO TC39 machine tools	ANSI/ASME B5 machine tools: components, elements, performance and equipment
Computer technology applied to manufacturing	ISO TC184/SC4 industrial data and manufacturing languages	ANSI/US PRO/IPO product data representation and exchange (IGES and STEP)

Library structure (ISO 13584, from Working Group 2 [10]), and Manufacturing Management Data representation (from Working Group 8).) The relationship between product data and process data, including manufacturing resource data, is yet to be explored within the standards committees. It is anticipated that the standardization of MR data at the international level will ultimately include participation and/or liaison by ISO TC29 (small tools), TC39 (machine tools) and TC184/SC4 (industrial data). A summary of relevant ISO and ANSI standards committees and their correlation is provided in Table 6.2.

The scope of ISO TC29 is defined as the standardization of hand tools and 'tools to be used on machines'. The focus of TC39 is 'standardization of all machine tools for the working of metal, wood, plastic, operating by removal of material or by pressure.' These technical committees are organized into numerous subcommittees and working groups to address a variety of subtopics (see Fig. 6.11) [12].

ISO TC29 - Small Tools		ISO TC39 - Machine Tools	
SC2 -	Drills, reamers, milling cutters and milling machine accessories	SC2 -	Test conditions for metal cutting machine tools
SC4 -	Screwing taps and dies	SC3 -	Modular units for machine tools
SC5 -	Grinding wheels and abrasives	SC4 -	Woodworking machines
SC8 -	Tools for pressing and moulding	SC6 -	Noise of machine tools
SC9 -	Tools with cutting edges made of hard cutting materials	SC8 -	Work holding spindles and chucks
SC10 -	Assembly tools for screws and nuts, pliers and nippers		

Fig. 6.11 Structure of ISO committees for cutting tools and machine tools.

In a similar fashion, ANSI committees are also divided into technical committees to address specific elements with the committee scope. The structure of the ANSI committees for cutting tools and machine tools is illustrated in Fig. 6.12 [20]. ANSI/ASME B94 and ANSI/CCPA B212 together address the technical subjects covered by ISO TC29. Although B94 is concerned with cutting tools, holders and other accessories in general, B212 focuses specifically on turning and boring tools with hard cutting edges of cemented carbide, oxide, and compacted diamond and their respective holders. Standards efforts of ANSI/ASME B5 include machine tools and their components (including work and tool holding elements) and performance tests for machine accuracy. Perishable tools are not within the scope of the B5 committee.

Although representatives from the various committees may have an interest in MR data and all of the committees may play a role in MR data standardization at some point in the future, current standardization efforts are active strictly within ISO TC29. (Standards efforts are dynamic

ANSI B94 - Cutting Tools	ANSI B212 - Carbide Tools	ANSI B5 - Machine Tools
TC3 - Basic nomenclature of cutting tools	TC1 - Cutting tools and holders: nomenclature,	TC11 - Chucks and chuck jaws
		TC30 - Power presses
TC4 - Tool life testing with single point tools	classification, sizes, tolerances, and	TC31 - Press brakes and shears
	identification	TC34 - Modular machine units
TC5 - Milling cutters		TC43 - Ball screws
TC6 - Carbide tipped milling cutter bodies	TC2 - Test methods	TC45 - Spindles noses and tool shanks for
	U.S. TAG to ISO TC29/SC9	
TC7 - Twist drills	U.S. Advisory Group to ISO	machining centers
TC8 - Jig bushings	TC29/WG34	TC46 - Milling and boring machines
TC9 - Reamers		
TC12 - Cut and ground thread taps		TC47 - Drilling machines
		TC48 - Turning machines
TC14 - Gear cutting tools		TC49 - Grinding machines
TC16 - Punches and dies		TC50 - Components
TC17 - Rotary slitters and shear knives		TC51 - Manufacturing systems and components
TC18 - Spade drills		TC52 - Machining centers
TC19 - Hack, band, and hole saws		TC53 - Machine tool test codes and symbols
U.S. TAG to ISO TC29		U.S. TAG to ISO TC39

Fig. 6.12   Structure of ANSI committees for cutting tools and machine tools.

in nature and are continually evolving. This information is current, as of publication.) This committee has had an informal task group studying the concept of computerized machining data exchange for a number of years. At the September 1995 meeting of the TC29 committee, participants voted to formally recognize a new working group (WG34) to address this subject. ISO TC29/WG34 has been tasked to 'define an information model for the digital representation of machining data in accordance with ISO 13584 Parts Library standard.' The resulting standard is intended to 'enable the free exchange of data concerning machining operations by chip removal between various computer systems' [13]. The scope statement of TC29/WG34 explicitly requires the collaboration with ISO TC184/ SC4/WG2 [10].

The TC29/WG34 committee commenced with active participation from seven countries. The initial efforts of WG34 primarily address the development of a reference hierarchy to structure cutting tool information and development of a detailed representation for turning tool data. Future efforts will address data elements for other types of cutting tools. Several contributions from various perspectives have been received and discussed by the committee. From the US, the NIST RRM intramural project has submitted their proposed data structure (including both the requirements specification tabular form and the EXPRESS representation) for consideration. Current indications are that the NIST proposal has been quite well-received within WG34. Portions of the NIST contribution pertaining to turning tools hierarchy and data representation have been selected as the initial baseline WG34 draft standard. Member countries are tasked to review this structure to ensure that individual requirements are met.

Within the US, representatives from NIST, Caterpillar Inc., and other industry counterparts have initiated efforts to form an ANSI-sponsored advisory group to coordinate US input to ISO TC29/WG34. This group has been organized under ANSI/CCPA B212 with representatives from US manufacturers, manufacturing software vendors, and tooling vendors. In addition, a liaison between ANSI/ASME B94 and ANSI/CCPA B212 committees appears necessary to support this advisory group.

The standardization of MR data for machine tool information remains an area for future work. The ISO TC39 and ANSI B5 technical committees for machine tool standards are aware of industry interest in the subject, but have yet to initiate active standardization projects.

6.5. OVERVIEW OF SIMILAR MR DATA REPRESENTATION EFFORTS

Issues regarding the representation of MR data have been addressed by several industry and research groups around the world. Interest in a standardized form of manufacturing resource data, particularly for cutting tools, is widespread and several potential representation forms exist. This section summarizes some of the primary efforts in MR data representation and presents other viewpoints for consideration in the solution of this industry need.

First of all, manufacturing organizations have created (and are continuing to create) company-specific tooling databases for storing MR information. In addition, manufacturing software developers are continuing to create proprietary MR data representations for use within commercial manufacturing software applications. As discussed throughout this chapter, this proliferation of proprietary MR representations is the current situation that standards efforts are trying to overcome. Nevertheless, the fact that companies continue to create these representations indicates that the need exists to use this type of manufacturing data. Even with the availability of a standardized MR representation, special situations for company-specific data structures will still arise and company decision-makers will still determine that other solutions are more suitable to their environment. All organizations will not recognize the benefits of a standardized MR data structure as presented in this chapter. Valid reasons may exist for these situations and decisions. This viewpoint must be understood and considered during development of standards for MR data to accommodate industry needs to the fullest extent.

Other US industry efforts have included development of information models of MR data to capture company requirements and to allow integration of specified software applications. In particular, staff from Boeing Corporation have developed an information model of cutting tools for machining [18], and Texas Instruments has created a repository of tooling information for use by multiple internal applications [7]. Participants of the Department of Energy TEAM consortium have envisioned a common MR database as a central component of their planned engineering environment [28]. Other manufacturing facilities have more than likely performed similar work.

In August of 1995 representatives from several large US manufacturing companies (i.e. Caterpillar, Ford Motor, General Motors, Chrysler) initiated discussions with representatives from several cutting tool vendors (i.e. Kennametal, Valenite, Sandvik, Carboloy) to propose the development of a common structure for electronic tooling catalogs. This group of manufacturers expressed a strong interest in obtaining tooling information in a common electronic format directly from the tooling vendors to support each manufacturer's diverse tooling information requirements. This group of manufacturers and cutting tool vendors investigated several current efforts, including those of the NIST RRM intramural project and the CIM GmbH organization of Aachen, Germany.

The CIM GmbH organization has developed a product called ToolBase for electronic tool data exchange [26]. The CIMSOURCE FormBase component of ToolBase is an electronic tool catalog with two main functions. First, it enables a search for all tool modules based on relevant characteristics. Second, it enables the transfer of tooling data directly to a manufacturer's applications (e.g. tool management, manufacturing process planning, NC programming). Several cutting tool vendors within Europe (i.e. Kennametal Hertel, Widia, Plansee Tizit, Sandvik Coromant) have agreed to provide CIM GmbH with product information using a prescribed format (i.e. ToolBase/STANDARD OpenBase). CIM GmbH organizes this vendor data into a database (i.e. ToolBase/DataBase) that can be accessed by the CIMSOURCE FormBase software. The resultant is a single CD-ROM sent to all ToolBase customers containing the cutting tool data from participating tooling vendors in a common format. This electronic tooling catalog greatly reduces the effort required by the manufacturing facilities for the maintenance of MR data. The perspective represented by this collaboration is that of electronic tooling catalogs. The information content is that which is typically provided by cutting tool vendors in their printed catalogs. Although this information satisfies the data needs of electronic catalogs, it may be insufficient for many CAD/CAM/CAE applications or other internal uses of the data. Representatives from the CIM GmbH organization have been active within ISO TC29/WG34, with their contributions based on the data structures used by ToolBase.

The TOOL company from France markets a software application for manufacturability analysis and tool wear prediction [14]. This software considers aspects such as depth of cut, feed rate, part materials, tool characteristics, and machine characteristics in its operation. Participants from this organization have been active within the ISO TC29/WG34 effort, primarily from the viewpoint of a representation for experimental or 'technological' data. This perspective places a separate set of requirements on the expected content of the resulting standard.

Participants from five European countries have applied for ESPRIT Program (EC) funding for the TOOLEX project to 'develop a dedicated open format for storage and exchange of tool data' [17]. Details of this proposed project are proprietary and the current status of its funding approval are not known. The focus of this proposal appears to be in support of ISO TC29/WG34 objectives.

In Sweden, Scania Corporation and Sandvik Coromant have developed EXPRESS-G representations of machine tool, cutting tool, and plant facility data [24]. Expectations from this group are that their work will be submitted and integrated with the STEP (ISO 10303) standard, specifically within Application Protocol 214 (Core Data for Automotive Mechanical Design Processes). These EXPRESS-G models were developed from a higher-level perspective than those of the NIST RRM intramural project. In addition, the standardization viewpoint of this group was focused more towards the ISO TC184/SC4 (i.e. manufacturing information) perspective and integration with the current STEP structure.

6.6 FUTURE OPPORTUNITIES

This chapter has presented the industrial need for a standardized representation of manufacturing resource data, a case study of one proposed solution to this problem, and the status of current MR data standardization efforts. Standardization efforts for representation of MR data have recently begun. Additionally, current standards efforts address only a limited subset of MR data types, namely cutting tools for machining operations.

Several future opportunities exist for interested parties to participate in research, development and/or standards activities. Industry feedback on current proposals and participation in standards committees is crucial to accurately capture industry requirements from various perspectives. The impact of standards efforts is directly related to the amount of industry and vendor involvement to ensure eventual use and implementation. Industry and vendor involvement ensures that not only are the correct problems are being addressed, but also that the proper issues are considered and acceptable solutions are developed.

Along with the level of interest exhibited by industry, a measure of success of standards work is the eventual use of the standard through implementations. The ultimate objective is the commercial implementation of a standardized MR data structure by manufacturing software developers and manufacturing resource providers. It is expected that future participation and advancements by these organizations will provide additional electronic commerce and software integration capabilities to the manufacturing industry.

Additional future efforts must address the extension of the current scope of standardization efforts to include other manufacturing resource types. The selected scope of current efforts is but a small subset of possible manufacturing resource areas. Other machine tool and tooling component categories exist, as well as other manufacturing operations (e.g. grinding, casting, welding, laser cutting, inspection) and other engineering, manufacturing and business functional areas within the product lifecycle (e.g. simulation, analysis, scheduling, accounting inventory). A broader class of software applications within the product lifecycle could also be addressed. It is expected that the discussion included in this chapter will be sufficient to stimulate interest in common representations of MR data and that future efforts will be initiated to address other manufacturing resource data types.

DISCLAIMER

REFERENCES

[1] ANSI/CAM-I 101 (1989) Dimensional measurement interface standard (DMIS), Consortium for Advanced Manufacturing, International (CAM-I), Arlington, TX.
[2] ANSI/EIA RS-274-D-1980 (1980) Interchangeable variable block data format for positioning, contouring, and contouring/positioning numerically controlled machines, Electronics Industries Association, Washington, D.C.
[3] ANSI/EIA RS-494-B-1992 (1992) 32-bit binary CL (BCL) and 7-bit ASCII CL (ACL) exchange input format for numerically controlled machines, Electronics Industries Association, Washington, D.C.

[4] ANSI X3.37–1987 (R1993) (1993) Programming language APT, American National Standards Institute, New York, NY.

[5] ANSI/US PRO/IPO 100–1993 (1993) Digital representation for communication of product definition data (revision and redesignation of ANSI/ASME Y14.26M), Initial Graphics Exchange Specification (IGES), Version 5.2, American National Standards Institute, New York.

[6] Barkmeyer, E. J., Hopp, T. H., Pratt, M. J., and Rinaudot, G. R., (eds) (1995) SIMA background study – requisite elements, rationale, and technology overview for the systems integration for manufacturing applications (SIMA) program, NISTIR 5662, National Institute of Standards and Technology, Gaithersburg, MD.

[7] Texas Instruments Corporation, (1988) Cutting tool management system, Requirements Specification, Internal Report, Defense Systems and Electronics Group, Dallas, TX.

[8] ISO 10303–1 (1994) Industrial automation systems and integration – product data representation and exchange – Part 1: overview and fundamental principles, International Organization for Standardization, Geneva, Switzerland.

[9] ISO 10303–11 (1994) Industrial automation systems and integration – product data representation and exchange – Part II: EXPRESS language reference manual, International Organization for Standardization, Geneva, Switzerland.

[10] ISO/CD 13584–1 (1995) Industrial automation systems and integration – parts library – part 1: overview and fundamental principles, TC 184/SC4 Document N289, International Organization for Standardization, Geneva, Switzerland.

[11] ISO/IEC 9506–2 (1990) Industrial automation systems – manufacturing message specification – Part 2: protocol specification, International Organization for Standardization, Geneva, Switzerland.

[12] ISO online (1996) (http://www.iso.ch), International Organization for Standardization, Geneva, Switzerland.

[13] ISO/TC29 N2218 (1995) Resolution 15, Stockholm 1995, Report of ISO/TC29/Task Group, computerized machining data exchange – creation of a new working group, International Organization for Standardization, Geneva, Switzerland.

[14] ISO/TC29/TG N23 (1994) Tool life tests, the method of T.O.O.L., International Organization for Standardization, Geneva, Switzerland.

[15] Jackson, K. (1993) U.S. project aims to chop new-car development time. *Automotive News*, May 31, 6.

[16] Jurrens, K. K., Fowler, J. E., and Algeo, M. E. A. (1995) Modeling of manufacturing resource information, requirements specification, NISTIR 5707, National Institute of Standards and Technology, Gaithersburg, MD.

[17] Kjellberg, T. (1995) Personal correspondence regarding draft TOOLEX project proposal, Royal Institute of Technology, Stockholm, Sweden.

[18] Boeing Corporation (1991) Machining cutter data model, Draft, Internal Specification, Seattle, WA.

[19] New program helps manufacturers slash lead time, Technology Trends. *American Machinist*, Nov. 1993, 14–16.

[20] ASME AS-11–1994 (1994) Personnel of codes, standards, and related accreditation and certification committees, The American Society of Mechanical Engineers, New York.

[21] Rapid Response Manufacturing (RRM) Consortium (1993) Detailed work plan for the rapid response manufacturing program, Preliminary Review Draft, National Center for Manufacturing Sciences, Ann Arbor, MI.

[22] Rapid Response Manufacturing (RRM) Consortium (1996) Steering committee minutes, Internal Program Document, National Center for Manufacturing Sciences, Ann Arbor, MI.

[23] Rapid Response Manufacturing (RRM) Intramural Project (1995) Program of the manufacturing engineering laboratory – 1995 – Infrastructural technology, measurements, and standards for the U.S. manufacturing industries, NISTIR 5599, National Institute of Standards and Technology, Gaithersburg, MD, 206–208.

[24] Scania Corporation (1995) Machine tool, cutting tool, and plant schemas, draft EXPRESS-G diagrams, Södertälje, Sweden.

[25] Sferro, P. R., Bolling, G. F., and Crawford R. H., (1993) Consortium tests DE. *Manufacturing Engineering*, June, 61.

[26] Sprung, M. (1995) ToolBase – The electronic tool data exchange, presentation slides, CIM GmbH, Aachen, Germany.

[27] Association for Manufacturing Technology (1994) Standards handbook: directory and guidelines for the use of standards in machine tool and related industries, First Edition, McLean, VA.

[28] Technologies Enabling Agile Manufacturing (TEAM) Consortium (1994) Strategic plan, Industry Review Draft, US Department of Energy, Oak Ridge, TN.

CHAPTER SEVEN

A group technology knowledge-based system for a rapid response manufacturing environment

Ali K. Kamrani and Peter R. Sferro

7.1 INTRODUCTION

The emergence of global markets for engineered products, and the resulting increase in competition in markets traditionally dominated by US manufacturers, has led to calls for increased productivity. Attention is particularly focused on understanding engineering design and developing new methodologies to increase the efficiency of the design process. One such method is concurrent engineering (CE). CE has focused on developing the tools and techniques for designing products. However, design in many industries is evolutionary, consisting primarily of incremental changes to existing products. This is known as the variant design engineering approach. Although concurrent engineering is concerned with integrating people with traditional engineering skills, the variant design engineering approach is concerned with empowering people with new skills by giving direct access to complete knowledge about design and manufacture of parts, with the philosophy that 'the knowledge is not generated at design time, but is retrieved from an engineering database which completely documents existing designs'.

To consider this issue, a rapid response manufacturing program was launched at the Ford Motor Company with 11 industrial partners. RRM is an integrated and computer-based environment which provides the engineer with direct access to complete knowledge and historical data regarding the design and manufacture of a product or similar product as it is being designed and developed. In this process known as direct engineering (DE), existing knowledge of the manufacturing processes of the product is documented electronically in the engineering database. A new product can be developed as a variation of an existing model. This will provide both manufacturability and a functionality analysis capability. This concept has revolutionized the engineering design approach.

This chapter will present the result of an ongoing research project on the development of a methodology for a group technology knowledge-based system which will support the associativity analysis required for integrated product design and process planning within the RRM environment. The developed methodology matches the criteria set by design for manufacture; the system will respond to any changes made to the design features and in turn will predict the process(es) feasible for the production of the product and the code(s) necessary for the retrieval of the data associated with the predicted process(es).

7.1.1 Rapid response manufacturing and direct engineering

The rapid response manufacturing (RRM) and direct engineering®(DE) concepts will provide an environment for a variant design process to use knowledge of a previously designed product in support of the development of new products. Automating the concurrent engineering philosophy within this environment will revolutionize the design of parts and their processes.

The main goals associated with developing such an environment is to integrate technologies such as product representation using feature-based modeling, knowledge-based applications to support product and process life cycle; an integrated and automated environment using distributed computing system; a direct manufacturing concept; and a concurrent engineering methodology. This environment requires a product modeling technology to support comprehensive knowledge of the design and fabrication of a product simultaneously. Feature-based modeling is used to represent this concept within this intelligent environment. This program is structured on a knowledge intensive environment available to the engineers at any time, using knowledge-based technologies to provide a decision support utility throughout the design life cycle. To support the integration and automation, RRM requires the use of state-of-the-art technologies in computer hardware, software and networking. The results of integrated product and process modeling and the knowledge-based application for improving design and manufacturing cycle time is justified by rapidly producing the part and its family. Direct manufacturing using freeform fabrication can be used to verify the geometric design representation of the product. Rapid response manufacturing should support an environment with the following goals and objectives.

Goal: product modeling based on the feature-based modeling approach and knowledge-based applications tools to support the entire product life cycle.

Objectives:

- provide direct and timely access to all related product design and manufacturing process knowledge throughout all phases of the life cycle; produce an accurate first part.
- support single-pass design based on direct trade-off optimization of the conflicting requirements imposed by manufacturability, reliability and functionality
- represent all design and process data using a single and complete model that drives all downstream applications.

Goal: an engineering environment built around object oriented and distributed computing system.

Objective: demonstrate the capability of the chosen computer architecture.

Goal: direct manufacturing techniques based on freeform fabrication and rapid prototyping.

Objectives:

- create and use a variant design process
- manufacture the part directly.

In summary, the goal of this program is the development of a knowledge-intensive environment in which engineers can design and manufacture parts faster, with acceptable levels of quality and cost. Figure 7.1 illustrates the process of engineering in this environment.

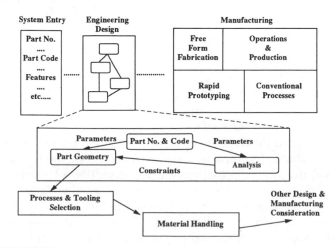

Fig. 7.1 RRM/DE environment.

7.2 TECHNICAL APPROACH

7.2.1 Geometric modeling and computer-aided design

Engineering design is a partial realization of the designer's concept. The designer can communicate with others using engineering drawings, and the information from these drawings can be used by the machinist for the manufacture of the product. As parts become more complex, designers will require a more sophisticated modeling approach and tools for design development and analysis. A computer-aided design system will provide the necessary environment for the designer to create, analyze, modify and optimize an engineering design [1]. CAD systems are categorized based on the system hardware (PC, work station, main frame), the application area (mechanical, architectural, etc.), and the modeling technique (2D, 3D, etc.). Figure 7.2 classifies the geometric modeling approaches.

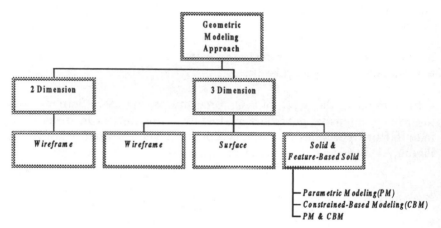

Fig. 7.2   Geometric models classification.

Geometric models are classified as both two- and three-dimensional. A 2D model is usually considered as a wireframe model. Wire frame models require less computing time and memory space, and will provide no information regarding the surfaces on the part, although they contain an accurate geometric description of the object being modeled. 3D geometric modeling includes wireframe, surface and solid modeling. Surface models will illustrate the mathematical description of the object being designed. They will also provide the ability to visually inspect the model in a 3D coordinate system. They are best used for the representation of complex surface contours (an automobile body). The concept of solid modeling is similar to the construction of an object as if it were actually being fabricated. The solid modeling approach is considered to be both iconic (mock-ups) and symbolic (mathematical descriptive) [2]. It is the most comprehensive approach and it contains the necessary information and data regarding the product's features which are stored in the design database. This information can be used for the development of a comprehensive knowledge-based system for product design purposes.

7.2.2 Feature-based design and modeling (FBM)

Feature terminology is used to define specific characteristics. There are many different definitions associated with the feature. From a designer point of view the feature can be defined as a specific design functionality, whereas from a manufacturing point of view a feature can illustrate a certain manufacturing process. In a feature-based design environment, features are used to illustrate associativity between both design and manufacturing using 'standard features'. Table 7.1 illustrates sample associativities between both design and manufacturing based on standard features.

In a feature-based design system, features are defined by a set of parameterized data [3]. This is known as the parametric design approach. In this approach the designer will define a set of geometric constraints and engineering relationships that are used to create the geometry of the object, and also to establish the associativities among the objects within the design itself. A set of expressions and variables is used to define the dimension of the object. When the numerical quantities of parameters are changed, the characteristics of the features are also updated concurrently. Although it is considered to be a complex approach, it will provide the

Table 7.1 Sample product design features and process machine associativities

Process features	Standard design features
Gun-drill machine	hole
Broach machine	flat face
Slotter machine	keyway
OD-grinder machine	outer cylinder
Turn-broach machine	flat face – outer cylinder
Induction heating machine	outer cylinder – fillet – chamfer
Draw furnace machine	hole – outer cylinder – fillet – keyway
Wire brush machine	hole – outer cylinder – fillet – chamfer – keyway
CNC lathe machine	hole – outer cylinder – fillet – chamfer – flat face
Shot blaster machine	hole – outer cylinder – fillet – chamfer – flat face – keyway

necessary flexibility and increased designer efficiency by creating a new design by altering existing models.

Parameterized features of mechanical parts are grouped into three classes. The first class contains standard dimensions. An example of the standard dimension is of a keyway (width, height and length) on a shaft. The second class, such as chamfer and formed shapes, which do not have standard dimensions, must be provided by the designer. Finally, the third class includes special and unique features associated with the part and its family (lightning holes on a crankshaft).

7.2.3 Geometric dimensioning and tolerancing (GD&T)

GD&T is a technique to dimension and tolerancing of a design with respect to the actual function or relationship of the part features. This will allow for a more efficient and economical approach of production. This is a major factor in controlling the quality of the part during the development of the process plan and also the production. The advantages associated with using the GD&T standard may include [4]: saving money by providing maximum producibility of a part through maximum production; ensuring that the design dimensional and tolerance requirements are carried out; and providing uniformity and standards to the design development and interpretation.

An integrated CAD/CAM environment requires standardization. GD&T can provide such a standardization and therefore has the ability to adapt to automation and computerization in an integrated design and manufacture. Table 7.2 lists the associativities between the GD&T standard and process machines. Tables 7.1 and 7.2 can be used to select processes and machines using design features and GD&T features.

Table 7.2 Sample product GD&T specified design and process machine associativities

Process machines	GD&T features for outer cylinder tolerances (mm)					
	Size	Tolerance	Rondns	Position	Finish (μm)	SRR
Grinder machine	0.5–1000	0.004–0.4	0.005–1000	0.05–1000	0.2	0.25–1.5
Turn-broach machine	0.2–100	0.17–0.47	0–7	0–7	1	0.127–1007
Induction heating machine	0.7–1007	0.17–0.47	0–7	0–7	1.7	0.127–1007
Draw furnace machine	0.7–1117	0.07–0.7	0–7	0–7	1.7	0.127–1007
CNC lathe machine	0.2–1000	0.04–1000	0.02–1000	0.1–1000	1	0.125–1000
Shot blaster machine	0.7–1007	0.27–0.87	0–7	0–7	1.7	0.127–1007

The features in GD&T are categorized into three groups. Individual features are those that are related to a geometric counterpart of itself and have no datum for reference. The datum are reference surfaces that are used to make part measurements. The related features are those defined as using one datum or several datum. The third category of features are those that could be considered as both individual or related. Table 7.3 lists the GD&T standards specified by ASME Y14.5M–1994.

Table 7.3   Standard GD&T classification

Geometric symbol category	Characteristics	Datum references
Form	Flatness Straightness Circularity Cylindericity	Never uses a datum reference
Orientation	Perpendicularity Angularity Paralleism	Always uses a datum
Location	Position Concentrity	
Runout	Circular runout Total runout	
Profile	Profile of line Profile of a surface	May use a datum reference

7.2.4 Group technology (GT)

Group technology is the realization that many problems are similar and therefore, by grouping similar problems, a single solution can be found. Group technology is generally considered as a manufacturing philosophy which identifies and exploits the sameness or similarity of parts and operation processes in the design and the manufacture of products. It has also been recognized that GT is an essential element of the foundation for the successful development and implementation of computer-integrated manufacturing through the application of the part family formation and its analysis.

To achieve higher productivity from design to manufacture, many manufacturing industries have become interested in the implementation of GT.

These companies apply their principles in their own way, although in some cases it is not identified as GT, but is considered simply as good managerial and operational practice [5,6]. A part classification system, which is considered an essential part of the grouping task, can be evolved as a means of describing parts which can be readily integrated with the database.

Grouping parts into families is a tedious task that requires careful planning and consideration. Basic methods that are available to solve the GT problem in manufacturing can be classified into: classifications, production-flow analysis, and cluster analysis [7]. Of the three proposed methods, coding and classification is the most efficient approach.

Classification is defined as a process of grouping parts into families based on some set of rules and principles. This approach can be further categorized into the visual method (Ocular) and coding procedure. Grouping based on the ocular method is a process of identifying part families by visually inspecting parts and assigning them to families and the production cells to which they belong. This approach is limited to parts with a large physical geometry and it is not an optimal approach because it lacks accuracy and sophistication. This approach becomes inefficient as the number of parts increases. The coding method of grouping is considered to be the most powerful and reliable. In this method, each part is inspected individually by means of its design and processing features.

Coding can be defined as a process of tagging parts with a set of symbols that will reflect the part's characteristics. A well-designed classification and coding system may result in several benefits for the manufacturing plant. A part's code can consist of a numerical, alphabetical or alpha-numerical string.

The attributes used for the development of a coding and classification system are categorized into three major classifications [7–9]. These include design, manufacturing, geometric dimensions and tolerances and geometric attributes. Design attributes are associated with the physical characteristics of the object. These may include the general shape of the product (rotational or non-rotational) and material (cast iron or steel). The manufacturing attributes of a product may include the steps required for the production. These may include numbers of basic processes, types of basic processes, and types of tools used. Dimension and tolerance is associated with the specification of the product. Dimension is required to convey complete information regarding the description of the detail of the part where tolerance is required to specify the acceptable range of variation in a dimension and shape.

7.2.5 Computer-aided process planning (CAPP)

The quality of the finished part is considered as one of the main issues in both design and manufacturing. For process planning purposes, the quality of a product during design is specified by means of desired surface finishes, maximum size tolerance, geometric dimensioning and geometric tolerance data.

The amount of knowledge required for the development of routing sheets and process planning is extensive. A considerable amount of knowledge must be extracted from an experienced process planner, classified, refined and finally formalized. The knowledge required for the development includes [10,11]:

- workpiece and surface features
- product quality
- machining operations
- machine tools
- tooling
- fixtures and dunnage
- features extraction capabilities
- operation selection
- sequence
  - operation constraints
  - geometric constraints
  - tooling constraints
- machine tool selection
- tool selection
- fixtures and dunnage selection
- machining parameters

Variant process planning is one of the techniques used in the computer-aided process planning approach. In the variant approach, the system takes advantage of the similarities among the components and retrieves an existing template process plan. This template (standard plan) is stored permanently in the database and can be accessed by establishing the similarity between the new design and the group of parts where the standard plan is developed. When the template plan is retrieved, a certain modification will be required to adapt for the new design. This task is performed either manually by the process engineer or a by knowledge-based system which recommends and makes the necessary modification.

7.3 THE GROUP TECHNOLOGY KNOWLEDGE-BASED SYSTEM METHODOLOGY AND ARCHITECTURE

Specification in an engineering design provides the technological data details of a product and are interpreted into process specification which are used for production. In a CAD environment, there are no capabilities to capture and define these data, although this information is critical to a computer-aided process planning (CAPP) environment. This lack of capability can be accommodated by incorporating a group technology coding system, which could be used for product design identification, CAD/CAPP integration and process prediction [12].

A methodology for the development of a group technology knowledge-based system for process planning based on GD&T standards and geometric features is proposed. The developed system assists the designers during the design of a new part to check the feasibility of designed features. The system will assist the designer to predict manufacturing codes based on the criteria systems set by design for manufacture and design for functionality methodologies. These codes are then used for the development of the route sheet for part production.

An extensive literature survey has revealed that very few approaches are aimed at the development of a methodology for coding and classification of parts into part families, based on both design and manufacturing attributes and their conformance with DFM methodology. Some researchers have applied DFA rules in identification of the necessary attributes for the GT system development. The primary reason is that there is no method accepted universally for the coding and classification of parts. It varies from one company or manufacturer to another. Thus the design and manufacturing attributes that work for a particular company may not be suitable for another company. Fig. 7.3 illustrates the structure of this coding system.

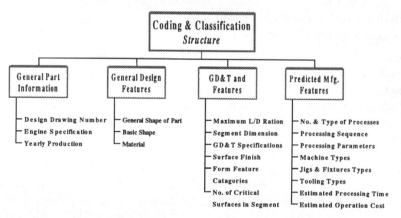

Fig. 7.3 The coding system structure.

7.3.1 The coding system module

Principles which were considered during the development of this coding methodology included [13]:

- utility: any classification system must be capable of filing and retrieving information easily, and be accurately based on the selected application

- efficiency: the system should provide a balance between coding, classification and retrieval
- attribute selection: the selected attributes should be based on ease of identification, significance and permanence
- relationship: the hierarchy of the coding and classification should progress from general to specific
- inclusiveness: the system should provide for grouping within a family of every item to be included
- Discrimination: the system should provide the user with comprehensive relationships and attributes that can aid in the selection of the desired attribute accurately
- Flexibility: accommodation of future expansion with minimum restructuring
- Standardization: recognize and accept industrial terminology.

The coding system is intended for part code assignment and is matched to the application and the task of crankshaft family production. Based on the requirements of the part the system consists of four major categories. The DCLASS (decision classification) software package is used for the development of rules and the decision trees. The objectives and attributes used for the development of this structure include the following.

General part information objective: To assist the designer in data filing, design retrieval and identification and costing.
Attributes:

- engineering design number: each design drawing has a number that is used for design identification
- engine specification: this code is used for engine type specification; the study focused on eight cylinder engines
- yearly production: the yearly production was also selected as a key factor due to its use in capacity and cost analysis.

General design features objective: perform family identification, complexity scope analysis, primary and secondary process prediction and costing.
Attributes:

- general shape of the part: the main physical shape of the part is defined using this attribute; the focus of this project is on rotational parts
- basic shape: further specifications and features of the crankshaft are defined
- material of the workpart: this attribute defines the material of the workpart; the material determines, to a great extent, the primary processes.

Geometric dimensioning and tolerancing features objective: to establish a structure for product features and quality specification, secondary and finishing process prediction.

Attributes:

- length/diameter ratio: the attribute is the ratio of the length of the workpart to the largest diameter of the workpart; it assumes a greater significance when the part is to be used in an assembly, and it also helps in the estimation of the size of the jig or the required fixture
- segment dimension: the crankshaft was divided into five major sections: post, counter weights, pins, main journal and flange; this part of the code provides the dimensions on each section
- geometric dimensioning and tolerancing: it is always difficult to produce a part within the exact dimensions and specification, therefore some degree of allowance must be given to the workpart. The tolerances allowed on the workparts are dependent on the application for which it is made
- surface finish: this attribute defines the finish condition of the surface
- standard form feature categories: the form features associated with each segment of the crank could be categorized as both concentric (grooves, center, inside diameter and face), and non-concentric (keyway, flat and face)
- number of critical surfaces: this section identifies the working (pin surfaces) and non-working (counterweight top) section of each segment

Predicted manufacturing features objective: establish an environment and a structured approach to allow for manufacturability and feasibility analysis, and intelligent process planning.
Attributes:

- number and type of processing steps: this attribute defines the number and type of the processing steps required to obtain the defined geometry
- processing sequence: this attribute provides the optimal sequence of the operations to be performed
- processing parameters: the performances of the machines and tools used, along with the requirements of surface finish, are determined by these parameters
- type of machine used for processing: this attribute provides information regarding the type of machine used for each of the operations; the machine chosen is based on manufacturability, availability and feasibility
- jigs and fixtures: this attribute provides information regarding the holding devices
- required tooling: this attribute informs the designer of the type of tooling required for the final geometry

- estimated processing time: this attribute defines the time required to perform all the required operations to obtain the final geometry
- estimated machine operation cost: this attribute gives the total machine operation cost.

7.3.2 Process planning module

The integrated process planning system is designed to retrieve the design input data from a text file. The engine component assembly file (ECAF) contains the parametric data associated with the design of a crankshaft solid model. The ARIES system is used for the development of the CAD model from this file. A set of rules is used for the verification and validation of the input data for process planning, which will prompt the user in the case of a missing parameter value. The main advantage of this system is that all the components can be defined as ECAF format and a set of rules can be used to check the data integrity and completeness for both CAD and CAPP. Figure 7.4 illustrates the overview structure of the proposed system and functions associated with each module. Fig. 7.5 illustrates an example of an ECAF template and the variables associated with the file. This file is imported using a user-defined FORTRAN routine linked to the DCLASS main module.

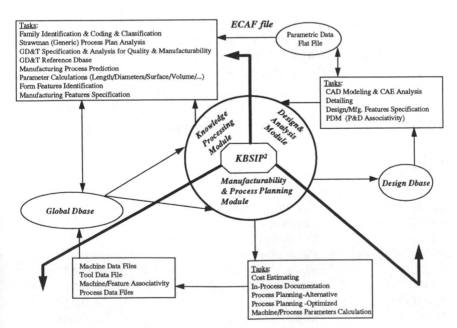

Fig. 7.4 Knowledge-based system structure for integrated product design and process planning.

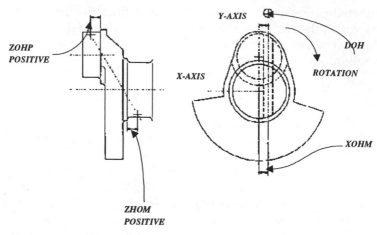

Fig. 7.5 Sample ECAF template file.

The GD&T data are retrieved by the system, either manually by user interaction or automatically from an EXCEL file. The retrieved data from EXCEL are reformatted into ASCII data (text), which are then imported into the KB system for analysis. This module is used mainly for manufacturability analysis and for process prediction. It is also used to search the part family archive to find similar operation plans. An example of this file is given in Table 7.4.

Several calculations are required for the determination of the removal volume. This information is then used to determine the number of surfaces which can be machined at one time, based on the machine

Table 7.4 GD&T data input file, ASCII format

Variable	Definition	Sample value
LPLT	Pin length tolerance	0.05
DPDT	Pin diameter tolerance	0.01
DPC	Pin diameter cylindericity	0.008
DPSURF	Pin surface finish	0.3
LM1LT	Main 1 length tolerance	0.4
DMDT	Main diameter tolerance	0.01
DM1RO	Main 1 diameter runout	0.25
DMC	Main cylindericity	0.008
DMSURF	Main surface finish	0.3

capabilities and horse power. As an example, the removal volume for the counterweight is calculated by

$$CWTMVOL(XCT).4 = (0.8 + TMS)*SA/360*3.14*((RCWM**2)$$
$$- ((DM/2**2)) \tag{1}$$

$$CWTTVOL(XCT).4 = TCH(XCT)*3.14*SA/360*(((RCWM(XCT)$$
$$+ 3)*2) - RCWM(XCT)*2) \tag{2}$$

$$CWTPVOL(XCT).4 = ((0.8 + TPS)*3.14)*(((SA/360)*RCWM**2)$$
$$+ ((RMB**2) - RPS**2))) \tag{3}$$

where XCT is the number of iteration (e.g. the number of working surfaces), CWTMVOL is the volume removal from the main side of the counterweights, CWTTVOL is the volume removal from the top side of the counterweight, CWRTPVOL is the volume removal from the pin side of the counterweight, TMS is the thickness of the main journal shoulder, SA is the sweep angle A, RCWM is the radius of the counterweight from centerline of main, DM is the main diameter, TCH is the thickness of cheek, TPS is the thickness of the pin shoulder, RMB is the radius of main boss, and RPS is the radius of pin shoulder.

The developed KB process planning system will provide the following features for the designer:

1. the ability to import ASCII design files into the KB environment
2. the rules associated with the analysis of the design data
3. the rules associated with GD&T data analysis
4. the routine required for the material removal calculation
5. the rules associated with process prediction and associativity analysis
6. the GT-database structure for supporting the developed rule-based system
7. the design structure (strawman) necessary for the development of the process plan.

Figure 7.6 is the flowchart of the template file for the development of the route sheet.

A sample route sheet follows on page 138.

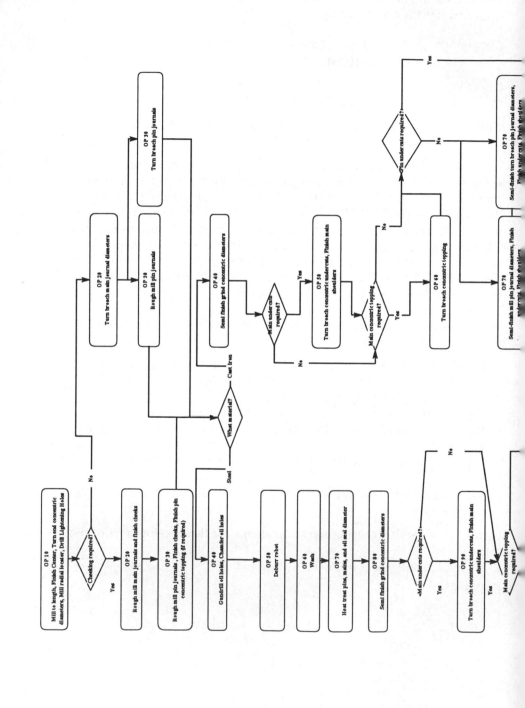

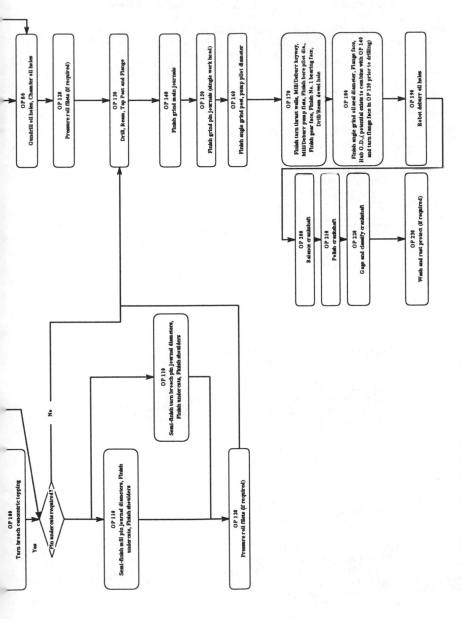

Fig. 7.6 Generic crankshaft process template.

Crankshaft Process Planning

Part ID: 6303   4.6L/pass/92   90
Part Name: 4.6L 90 deg V-8   Date: 03/12/1996
Planner Name: Kamrani   Time: 16:49
Annual Volume is 550000 at 48 Wks./Yr. and 120 Hrs./Wk. at .7 Efficiency
Net Pcs/Hr = 95., Gross Pcs/Hr = 127, Max Pcs/Yr = 733333, Material is: WSE-M1A17
Cheeking Required: YES   Topping Required: NO   Undercuts: YES

Oper. No.   *Process Description*

---

10   Step 1. ROUGH MILL CRANK TO LENGTH OF 567.43 mm.
             Post End material removal volume = 1227.9 cu.mm.
             Post End material removal rate = 2605.3 cu.mm./min.
             Hub End material removal volume = 2403.1 cu.mm.
             Hub End material removal rate = 5098.5 cu.mm./min.
             Rough cut depth for Post End and Hub End = 2.55 mm.
         2. DRILL CENTER (14 gage ball)
         3. MILL RADIAL LOCATOR
         4. LIGHTENING HOLES ARE REQUIRED
             Diameter of holes = 21.4 mm.
             Length of holes = 51.64 mm.
             Drill angle of holes = 18 degrees.
         Machines required = 2 @ 120 pcs/hr Cost = $6426000
         Special Tools = $124000 Perishable Tools = $30000
Note: Crank rough diameter = 188.00 mm.

---

20   Step 1. ROUGH MILL MAIN JOURNALS
             Material removal volume = 61841.70 cu.mm. for all Mains.
             Material removal rate = 131207 cu.mm./min.
             Depth of cut = 2.55 mm.
             Main diameter after cut = 67.943 mm.
         2. FINISH MAIN COUNTERWEIGHT CHEEKS
             Material removal volume = 95863. cu.mm. for all Cheeks.
             Material removal rate = 203391 cu.mm./min.
             Depth of cut = .4 mm.
         Machines required = 3 @ 60 pcs/hr Cost = $4869000
         Special Tools = $795000 Perishable Tools = $45000

---

30   Step 1. ROUGH MILL PIN JOURNALS
             Material removal volume = 84207.04 cu.mm. for all Pins.
             Material removal rate = 178659. cu.mm./min. for all Pins.
             Depth of cut = 2.55 mm.
         2. FINISH PIN COUNTERWEIGHT CHEEKS
             Material removal volume = 109151 cu.mm. for all Cheeks.
             Material removal rate = 231583. cu.mm./min. for all Cheeks.
             Depth of cut = .4 mm.
             Depth of cut = 2.30 mm.
         3. FINISH PIN CONCENTRIC TOPPING
             Depth of cut = 1.55 mm.
         Machines required = 4 @ 36 pcs/hr Cost = $6012000
         Special Tools = $596000 Perishable Tools = $ 28000

Oper. No.   *Process Description*

---

40 Step 1. SEMI-FINISH GRIND CONCENTRIC PIN DIAMETERS
            Material removal volume = 1599.76 cu.mm. for all Pins.
            Material removal rate = 3394.15 cu.mm./min. for all Pins.
            Finished diameter = 52.99 mm.
            Grind depth of cut = .05 mm.
        Machines required = 2 @ 80 pcs/hr Cost = $11132000
        Special Tools = $ 242000 Perishable Tools = $2066000

---

50 Step 1. TURN BROACH CONCENTRIC PIN UNDERCUTS
            Undercut of pin fillet = .4915 mm.
            Tolerance of pin undercut = +/− .06 mm.
            Length of undercut = 2.3 mm.
            Tolerance of length = +/− .1160 mm.
        2. FINISH MAIN SHOULDERS
        Machines required = 3 @ 62 pcs/hr Cost = $2265000
        Special Tools = $852000 Perishable Tools = $36000

---

60 Step 1. SEMIFINISH MILL PIN JOURNALS
            Material removal volume = 12852.88 cu.mm. for all Pins.
            Semifinish Pin diameter = 53.043 mm.
            Depth of cut = .40 mm.
        2. FINISH UNDERCUTS
        3. FINISH SHOULDERS
        Machines required = 3 @ 48 pcs/hr Cost = $4734000
        Special Tools = $ 429000 Perishable Tools = $45000

---

70 Step 1. PRESSURE ROLL FILLETS
        Machines required = 2 @ 120 pcs/hr Cost = $3202000
        Special Tools = $ 232000 Perishable Tools = $160000

---

80 Step 1. DRILL & REAM POST FLANGE
        2. TAP POST AND FLANGE
        Machines required = 2 @ 120 pcs/hr Cost = $7698000
        Special Tools = $ 286000 Perishable Tools = $48000

7.4 CONCLUSION

The variant design concept is concerned with empowering people with new skills by providing an environment for the direct access to complete knowledge about design and manufacturing of products as the new products are being designed. The developed system is based on the variant design approach, and it clearly illustrates the major role that a group technology coding and classification system can take on within an integrated product design and process planning environment.

REFERENCES

[1] Chang, T., Wysk, R. A., and Wang, H. (1991) *Computer-Aided Manufacturing* Prentice Hall, International series in Industrial and System Engineering, W. Fabrycky and J. Mize (eds).

[2] Amirouche, F. (1993) *Computer-Aided Design and Manufacturing*, Prentice Hall, Englewood Cliffs.

[3] Shah, J. J. (1992) *Features in Design and Manufacturing*. Intelligent Design and Manufacturing, A. Kusiak (ed.), John Wiley.

[4] Foster, L. W. (1994) Geo-metrics III: The Application Geometric Dimensioning and Tolerancing Techniques. Addison-Wesley.

[5] Kamrani, A., and Parsaei, H. (1994) A group technology-based methodology for machine cell formation in a computer integrated manufacturing. *Computers and Industrial Engineering International Journal.*

[6] Kamrani, A. K., and Parsaei H. R. (1994) A methodology for the design of manufacturing systems using group technology. *International Journal of Production Planning and Control.*

[7] Kusiak, A. (1990) *Intelligent Manufacturing Systems*, International Series in Industrial and Systems Engineering. Prentice Hall, Englewood Cliffs, New Jersey.

[8] Requicha, A. A. G., and Vandenbrande, J. (1988) *Automated Systems for Process Planning and Part Programming*, Artificial Intelligence: Implications for CIM, A. Kusiak (ed.), IFS Publications, Kempston, UK, 301–326.

[9] Senath, P. H. A., and Sokal, R. R. (1973) *Numerical Taxonomy: The Principles and Practice of Numerical Classifications.* Freeman Press, San Francisco.

[10] Chang, T. C., and Wysk, R. A. (1985) *An Introduction to Automated Process Planning Systems.* Prentice Hall, Englewood Cliffs, New Jersey.

[11] Zhang, H. C. and Alting, L. (1994) *Computerized Manufacturing Process Planning Systems.* Chapman & Hall.

[12] Kamrani, A. K. Agarwal, A., and Parsaei, H. (1994) *Automated Coding and Classifications System with the Supporting Databases for Effective Design of Integrated Manufacturing Systems.* The special issue of Intelligent Manufacturing Systems, Environmental and Intelligent Manufacturing Systems Series, Prentice Hall.

[13] Snead, C. S. (1989) *Group Technology, foundation for Competitive Manufacturing.* Van Nostrand Reinhold, New York.

CHAPTER EIGHT

Virtual prototyping in simultaneous product/process design for disassembly

Matthew D. Bauer, Zahed Siddique and David W. Rosen

Designers are being called on to design higher-quality products, to consider additional life cycle concerns, and do this more rapidly with fewer resources. Our approach to rapid product development centers on two main ideas: an intelligent partition of responsibilities between designer and computer based on their abilities and capabilities; and an integrated decision support environment in which all the design requirements can be considered simultaneously. We illustrate the application of these ideas to product design for disassembly through our approach to virtual prototyping and the formulation and solution of simultaneous product and disassembly process design problems.

Requirements on product life cycle concerns, such as service and recycling, during design raise the need to ascertain the product assembly and disassembly characteristics before physical prototypes are available. Virtual prototypes are proposed to support the assessment of assemblability/disassemblability by enabling designers to interact virtually with products, as CAD models, with guidance provided by a process reasoning module in a CAD system. Moreover, simultaneous product and process design is enabled with the introduction of virtual prototyping into the product development process. The emphasis in this chapter is on the rapid development of product disassembly processes via virtual prototyping, and the integration of product and disassembly process design in a decision support environment. Methods for automatically generating disassembly directions and tool changes, as well as capturing designer disassembly actions, are outlined in this work. The subsequent integration of these disassembly processes into product/process decision support problems is also described, as is the solution mechanism for simultaneous product/process design problems. By defining a design problem suitably in terms of product and process variables, constraints and goals, a multiobjective optimization code facilitates rapid and simultaneous improvement of product and process designs. To illustrate simultaneous product/process design, application of the method to the design of automotive interior components and their disassembly processes is presented. The results to date are summarized, and they indicate that virtual prototyping effectively enables rapid disassembly process design.

8.1 OUR FRAME OF REFERENCE FOR SIMULTANEOUS PRODUCT/PROCESS DESIGN

In recent years, the concept of green engineering has established a foothold in the engineering community. As a result, concerns about the environment have spurred interest in design for the life cycle (DFLC). Moreover, modern product development must be performed in the context of limited natural resources, i.e. not only should a product consume the minimum of resources during its useful life, but it should also preferably lead into another cycle of use either by material recycling or component reuse, or, at the very least, it should be environmentally friendly in disposal. Product design for service and demanufacture, of which disassembly is a principal component, is a necessary part of product development and will become even more important as the US follows the lead of European governments and companies. For instance, automotive service and remanufacture is rapidly becoming a critical design issue in light of the German car take-back initiative.

Due to these environmental concerns, design-for-disassembly (DFD) and design-for-service (DFS) guidelines have been (and continue to be) formulated to aid designers and engineers in reducing their product's end-of-life environmental impact. The typical approach in designing for X (DFX, X = manufacturing, disassembly, service, etc.) is to compile lists of general rules and then encode them in spreadsheets for the analysis of detailed product designs. Although a reduction in environmental impact can be realized through these approaches, they are not ideal, and problems have arisen with their use. The most serious drawback to these approaches is the need for detailed product models (preferably physical prototypes) with which to perform analyses. Accordingly, these approaches are generally utilized in the later stages of design as secondary design tools. Thus, these tools are never fully integrated into the design process, particularly in the early stages of design where the most significant impacts could be realized.

In this research, our goal is to develop a methodological foundation to support designers in assessing and improving a product's demanufacturability, emphasizing disassemblability, in the early stages of the product development process. (Demanufacturing characterizes the entire process involved in recycling, reuse, incineration, and/or disposal of a product after it has been taken back by a company.) That is, rather than redesigning a product to improve its disassemblability, DFD will be integrated into decision-making in the early stages of the design to ensure the development of conceptual designs that truly embody the principles of DFD, thereby realizing the benefits of DFD analysis while reducing time to market, minimizing secondary cost, maintaining product quality and performance and improving demanufacturability. Our immediate research goal is to develop the methodology and tools to support simultaneous product and disassembly process development via virtual prototyping in preliminary design. Our current work enables disassembly process simulation and analysis to be performed directly on product models, thereby decreasing the reliance on general design rules, eliminating product model re-entry, and increasing the fidelity of the analyses. Simultaneous product/process design is facilitated by the development of virtual prototypes and virtual prototyping capabilities which support the solution of parametric design problems, goal-directed geometry problems.

In addition, the work described in this paper is part of a project to develop a virtual design, service, and demanufacture studio (Rosen *et al.*, 1995). Currently, the virtual environment, which has three-dimensional graphics and a simple keyboard–mouse interface, supports the disassembly of product models. Product disassembly processes (e.g. disassembly sequences, tool changes, component disassembly paths, etc.) are generated in part by automated reasoning methods and completed by designers' activities performed on the virtual prototypes. The resulting processes become primary elements in simultaneous product/process design.

Our approach to design for disassembly via virtual prototyping is detailed in this chapter. In this chapter, we are more concerned with the presentation of our approach to design for disassembly via virtual prototyping than the detailed results of our example *per se*.

8.2 RELEVANT BACKGROUND

A brief review of existing research that supports our current efforts in simultaneous product/process design is presented in this section. Topics of discussion include designing for X and the generation of disassembly processes from CAD models. Then our approach to simultaneous product/process design via virtual prototyping is described.

8.2.1 Designing for X (X: assembly, disassembly and service)

The typical approach in design-for-X is to compile lists of rules and then encode them in spreadsheets for analysis of detailed product designs. For example, Beitz (1993) has developed tables to keep track of the number of parts, component materials, access to parts and fasteners, tools needed for disassembly, and the time to perform disassembly operations (among other measures) to evaluate and redesign assemblies in a manner similar to that of Boothroyd and Dewhurst (1991). DFD strategies include reducing the number of components, making use of fastening methods that allow quick and easy component separation, and eliminating the need to separate parts. Design-for-service (DFS) combines aspects of DFA and DFD in that if a component in an assembly needs to be replaced, then that component, as well as each that stands in its way, needs to be disassembled and subsequently reassembled. In addition to the ease of assembly and disassembly of each component, the frequency of repair/ replacement of service-prone components is considered (Marks *et al.*, 1993). Quantitative analysis schemes such as these are best suited to analysis of existing designs but fall short in suggesting the actual improvements to be made (Kroll *et al.*, 1988).

8.2.2 Generating disassembly processes from CAD models

The generation of disassembly processes can be divided into two main areas: the development of disassembly sequence, and the subsequent development of disassembly paths. The two basic approaches to the generation of disassembly sequences are interactive and automatic. Interactive approaches require substantial user knowledge and input but can accommodate complicated assemblies. On the other hand, automatic techniques can generate disassembly sequences directly from CAD product representations (such as boundary representations); however, the component and assembly complexity is necessarily low due to the computational strain imposed by the abundance of geometric and topological information and reasoning that is required. Huang (1993) utilizes an iterative process to generate a disassembly sequence at each stage of the assembly state, updating the model each time a part is removed. De Mello and Sanderson (1988) use an AND/OR graph to determine disassembly sequences. Lee and Shin (1990) use a liaison graph representation of an assembly to automatically determine preferred subassemblies, evaluate these subassemblies, and construct a dis-assembly tree. As mentioned above, automated disassembly sequence generation techniques are restricted to much simpler assemblies, because extensive geometric and topological information and reasoning are needed to determine the directions of feasible disassembly. Hoffman (1990) presents an approach that deals with the boundary representation of models, named the B-Rep Assembly Engine (BRAEN). Woo and Dutta (1991) describe a technique to construct disassembly trees from CAD models.

Once a feasible disassembly sequence is determined, disassembly paths can be formulated and 'optimized' to complete the disassembly process model. Although research in component disassembly path planning is scarce, much research has been performed in the area of collision-free path planning for mobile robots and automated guided vehicles (i.e. navigating objects around obstacles). Three approaches to the generation of collision-free paths are apparent in the literature. Two of these approaches have emerged from research in the field of computational geometry. The first is based on a configuration space that is defined by shrinking a moving object to a point and expanding obstacles by an equivalent proportion. Then the boundaries of the obstacles represent collision-free paths and the problem reduces to graph searching for an optimal path. The second approach is characterized by a representation of the free space between obstacles, in which the moving object is typically guided along the central lines of the free space primitives. The third approach, which is the one taken in this work, is to formulate path planning as an optimization problem. Although the first two methods may be more computationally efficient than the third, they are difficult to extend to three dimensions. In addition, they are often constrained to dealing with convex objects (Donald, 1987). Although the third approach is more easily extended to three dimensions, path planning optimization problems are difficult to formulate and implement and are typically quite expensive computationally.

Virtual prototyping (VP) is an emerging technology which has the potential to enable the simultaneous definition of disassembly process sequences and paths in an interactive virtual environment. Virtual reality is regarded as a natural extension of 3D computer graphics with advanced input and output devices (Connacher *et al.*, 1995). Virtual prototyping is the simulation of prototype designs in a virtual environment, and consequently it allows a designer to make important analyses of a design early in the design process. The requirements for a virtual prototype depend on the characteristics of the physical prototypes desired and on the type of analysis that will be performed. Graphical presentation and interaction with the product data model are two of the important factors involved in virtual prototyping (Dai and Gobel, 1994). Recent work in virtual prototyping for assembly includes that of Kuehne and Oliver (1995) and Connacher *et al.* (1995).

8.3 DESIGNING FOR DISASSEMBLY VIA VIRTUAL PROTOTYPES

The current state of design-for-disassembly is that of redesigning a product to improve its disassemblability, rather than designing dis-assemblability into the product. We propose simultaneous product and disassembly process design via virtual prototyping to support the development of conceptual designs that truly embody the principles of DFD, thereby realizing the benefit of DFD analysis without the undesirable expense often associated with redesign. In this work, disassembly process design is executed directly on product models through the evaluation of disassembly process constraints and goals via virtual prototyping, thus decreasing our reliance on general design rules, eliminating product model re-entry, and increasing the fidelity of our analyses. Moreover, results from simultaneous product/process design, such as disassembly path distances, may be incorporated into traditional DFD analysis tools to improve the fidelity of assessment. The principal steps to simultaneous product/process design are outlined in Fig. 8.1.

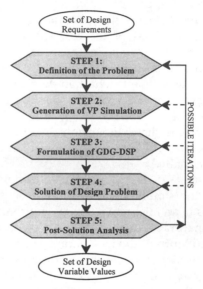

Fig. 8.1 Steps in simultaneous product/process design.

The first step in simultaneous product/process design is the definition of the problem. From a set of design requirements, a specific problem must be developed and expressed in the form of a problem statement. This problem statement should articulate the specific problem to be addressed in addition to the product and process requirements in the form of demands, wishes and preferences. The second step consists of the generation of the virtual prototype simulations. From a product model in CAD, a process model is developed in a virtual prototyping environment. By capturing disassembly motions in the virtual prototyping environment, a feasible (but not optimal) disassembly process is developed. This process is then animated to form the virtual prototype simulations. The third step is the formulation of a goal-directed geometry decision support problem (gdg-DSP) which is a mathematical representation of the designer's demands, wishes and preferences articulated in the problem statement. The fourth step is the solution of the gdg-DSP. Once the gdg-DSP is implemented on a computer, the virtual prototype simulations are used to evaluate appropriate product/process constraints and goals. Optimization techniques are applied to facilitate rapid exploration of the design space. Lastly, the fifth step in simultaneous product/process design is post-solution analysis. As illustrated in Fig. 8.1, the procedure may be iterative.

The system architecture which supports simultaneous product and disassembly process design is illustrated in Fig. 8.2. As shown in the figure, the five principal constituents of the simultaneous product/process design are: virtual prototyping, CAD, product representations, goal-directed geometry (GDG) and a solution algorithm.

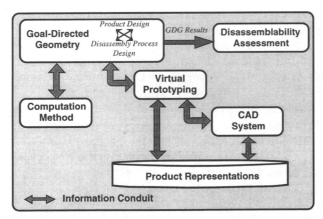

Fig. 8.2 Software architecture for simultaneous product/process design.

In the remainder of this section, three of the five constituents are detailed, namely, virtual prototype generation and virtual prototype simulation, goal-directed geometry, and optimization method for solution of simultaneous product/process design.

8.3.1 Generation of virtual prototypes for simultaneous product/process design

A virtual prototype is a model of a product and a process that the product undergoes. The key features of a virtual prototype for disassembly are illustrated in Fig. 8.3, along with a prototype of the rear module of an automotive center console being disassembled by a virtual disassembler with a screwdriver. The product model of a virtual prototype contains geometric component models and assembly information, including a hierarchy, mating relationships among components, fasteners and other

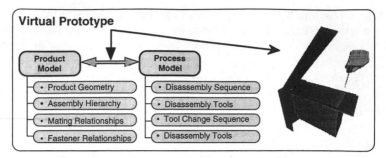

Fig. 8.3 Illustration of virtual prototype in SPDP design.

spatial relationships. Information contained in the process model varies with the type of process. For disassembly, the process model contains information about operation sequences, tool changes, component disassembly paths, etc. Elements of the process model can be generated automatically but the completion of the process model requires human interaction and assistance. Generation of a product's disassembly process through virtual prototyping can be implemented throughout the product realization process, from preliminary design to detail design, and the resulting virtual prototypes can be built upon as design progresses.

Our overall understanding of virtual prototyping and the design context in which it occurs is summarized in Fig. 8.4. Prototype generation and process simulation are the main activities involved in virtual prototyping. Virtual prototype generation consists of three steps: VP preparation in which product CAD models are converted into a representation suitable for prototyping, VP process design in which automated reasoning methods determine certain attributes of the process model, and VP maturation in which designers interact with the virtual prototype to fine-tune the process model. Lastly, process simulation is the

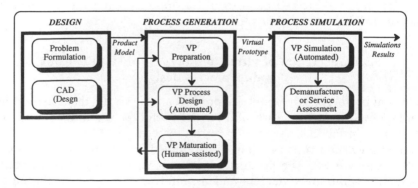

Fig. 8.4 Principal activities in virtual prototyping.

Table 8.1 Process model information and their acquisition method

Information	Manual methods	Automated methods
Component removal sequence	Resolve ambiguities	Partial sequencing
Fastener removal sequence	Resolve ambiguities	Identify fasteners
Tools		Select appropriate tools
Tool change sequence	Resolve ambiguities	Automatic after component and fastner sequences are known
Fastener accessibility and visibility	Global accessibility and visibility	
Access and removal paths	Global motions	Local motions

execution of the process model on the VP's product model to assess the design which is described below.

Referring to Fig. 8.4, virtual prototyping has as an input a product's CAD model and designer interactions, and its output must be suitable for disassembly simulation and assessment. Consequently, these activities drive the requirements of the virtual prototype representation and VP capabilities. The types of information that comprise a disassembly process model and their methods of acquisition are articulated in Table 8.1. As shown in Table 8.1, the generation of a disassembly process includes not only the definition of disassembly sequences but also the definition of tool change sequences, disassembly motions, and component and fastener removal paths. In the remainder of this section, the three principal steps to virtual prototype generation are detailed.

Virtual prototype preparation

Generation of a virtual prototype's product model involves the conversion of a product model in CAD (such as CODA) into a format which facilitates disassembly. (CODA (configuration design of assemblies) is a prototypical CAD system under development in the Systems Realization

Laboratory at the Georgia Institute of Technology.) In a nutshell, component graphics models and a simplified assembly model constitute the product model of a virtual prototype. Graphical component models are composed of polygonal surfaces, with material and mass properties which are used for broader demanufacturing processes.

The assembly model consists of a hierarchical assembly-subassembly structure, mating relationships, and fastener information. This information is acquired from CODA's product models and is restructured to facilitate disassembly reasoning and user interaction. In CODA, mating relationships are based on two fundamental relationships, fits and against, where the fits condition constrains the center-lines of two objects to be collinear, and the against relationship constrains two surfaces to be coplanar. Mating relationships are represented at two levels of detail in the CODA: between components and between component surfaces or handles. Handles are convenient abstractions of shape, such as center-lines or mid-planes, that are abstracted from a component's geometric model as illustrated in Fig. 8.5. Fastening relationships are modeled using

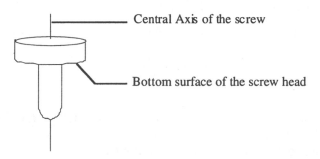

Fig. 8.5 Screw's mating handles.

composite mating relationships, i.e. constructed from two or more fits and/or against conditions. For example, a snap-fit would be modeled by one against condition and one fits condition. Fasteners are treated as special agents of composite mating relationships but have geometric and other properties like other components. Thus, a screwed-joint is modeled by two against and two fits conditions between the screw and two components. Kinematic joints, such as revolute and prismatic, are also modeled using composite mating relationships. A more detailed discussion of CODA's mating relationship representation and comparisons with other approaches is given by Rosen *et al.* (1996) and Hassenzahl (1994).

To facilitate disassembly process generation and user interactions, a library of standard fasteners is maintained in the studio. The information associated with each fastener includes: the type of tool needed for disassembly, the required unfastening motions, and the relative spatial transformation between fasteners and their corresponding tool. In the library, fasteners are represented as parametrized geometric models with predefined handles for use in mating relationships.

From the mating relationships in the virtual prototype, a list of fasteners and fastening relationships for each component is generated, that specifies which fasteners and fastener relationships must be unfastened for that component to be removed from the assembly. After the product model for a prototype has been defined, the next step is to generate a disassembly process for the product.

Virtual prototype process design

After completing the conversion of a CAD model into a virtual prototype, the prototype is disassembled by the designer in a virtual environment to determine a feasible disassembly process. The capabilities presented in Table 8.1 which are available in the studio to support disassembly are described in this section. The fastener removal sequencing is simple because each component maintains its own fastener list. Additionally, tool change sequencing is also simple because each fastener type is associated with a specific tool type. The selection of appropriate tools is accomplished through the use of a look-up table.

The determination of component disassembly sequences from CAD models is a difficult problem which has been the subject of much research. Definition of disassembly sequences depends on both local and global motions. Local motions denote infinitesimally small motions starting from a component's assembled position, whereas motions that depend on the presence of obstacles or other constraints are characterized as global motions. The determination of local motions is a two-step process in this work: determine local motions based on mating faces, and eliminate those motions which violate the fits conditions. What follows is a discussion of the methods used to determine local motions, and hence partial disassembly sequences, in our virtual prototyping environment.

The algorithm used to determine local motions for components is an extension of the technique introduced by Woo and Dutta (1991). The class of problems that are considered in Woo and Dutta's work is one-disassemblable (a single translation completely removes the component from the rest of the assembly), and it include three degrees of freedom (purely translation). The local motions are first determined by considering only mating surfaces (against conditions) then by testing the fits conditions, ignoring fasteners. The mating surfaces are determined from each component's mating relationships. Conceptually, if the normal vectors of all mating surfaces on a component are within 180 degrees of one another, then that component is movable locally. This property is referred to as monotonicity.

Monotonicity is determined by testing for the hemisphericity of mating surface normal vectors. In this work, the algorithm of Chen and Woo (1992) is used to determine hemisphericity. Basically, a component is locally movable if all mating surface normal vectors pierce only a single hemisphere of a unit sphere. Actually, the locus of disassembly directions is typically larger than just the pole. A generalized cone representation, similar to that used by Zussman *et al.* (1991), is used to represent feasible motion directions. By comparing the hemisphericity results with the constraints resulting from fits conditions, a component's local motions are calculated. Specifically, if a fits axis lies within the range of motion specified by the hemisphericity results, then that component is movable in the direction of that axis. All such axes on a component must be parallel or the component is not movable. The methodology in step two is an extension of the Woo and Dutta method. However, this algorithm does not provide global directions of removal for components. The local directions are calculated to assist the designer in completing the disassembly process.

Based on the results of local motion analysis, partial orders can be applied to component disassembly processes. During virtual disassembly, local motions are determined for all components remaining in the assembly. Those components that have some local motions are candidates for disassembly at that time. It is up to the designer to specify which component will be disassembled and in which order within the confines of the local motion analysis. Totally ordering disassembly sequences is a difficult problem that often involves factors that are not easily quantified. In the next section, our approach to completing the disassembly sequence is provided.

Maturation of virtual prototypes

To mature the disassembly process, the prototype created from the product model is disassembled in the studio. To gather meaningful and insightful information, disassembly activities performed in the studio need to mimic reality as closely as possible. Our approach to modeling the designers and their interactions with tools and components is presented in the remainder of this section.

A designer's presence in the virtual prototyping environment is represented by a virtual hand. To make the movements of the hand realistic, the hand must have the capabilities to grab tools, perform unfastening motions, and remove unfastened components or fasteners in a manner that is natural, easily describable and intuitive to the designer. Information about the position of the hand is necessary to perform these activities. In our approach, a reference point is identified on the hand which is used to store the orientation and position of the hand. This point is used not only to track the movement of the hand, but also as a reference point for grabbing tools and components.

The behaviors of tools (such as screwdriver, wrench, etc.) must also be modeled as accurately as possible to facilitate their use by the designer. Information about the location of the active end of the tool and the primary axes of unfastening motions are required to imitate the functions of the tools. In addition, a reference point and orientation of the local axes are required to correctly position the tool in the virtual hand. These points and axes are highlighted on the screwdriver illustrated in Fig. 8.6. Fortunately, the designer need not be aware of these coordinate systems.

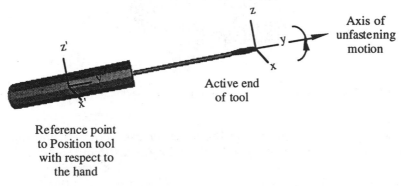

Fig. 8.6   Illustration of screwdriver in virtual environment.

Tools can be viewed as an interface between the hand and a fastener; as a result, two types of positioning are important: positioning of the tool with respect to the hand, and positioning of the tool with respect to a fastener. Both types of positioning have been semi-automated in the studio using an approach similar to that used in calculating end-effecter positions, relative to a workpiece, in robotics (Paul, 1981). A simulation module that incrementally changes the current position of the tool to the desired position is used to show the designer the path that was taken by the tool and hand. A bounding box method is used to check proximity between the tool and the fastener.

Fastener accessibility needs to be determined to evaluate a design for ease of disassembly. To help the designer in the studio, a simulation module has been created to mimic unfastening motions. The tool and the hand are simulated by rotating or translating them along the axis of unfastening motion of the tool. As an example, a screwdriver and hand are rotated along the screwdriver's axis to unfasten a screw. To activate the simulation, the tool's active end must be positioned and oriented correctly with respect to the fastener. From the simulation, accessibility and the relative difficulty of unfastening can be assessed by the designer. In Fig. 8.7 the unfastening motions required to unfasten a screw are shown.

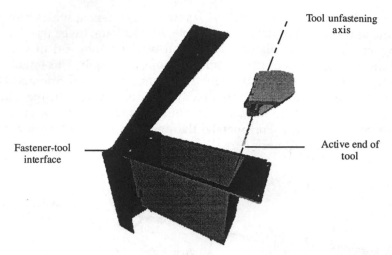

Fig. 8.7 Illustration of virtual disassembly.

The sequence of motions that the designer uses to access and remove fasteners and components is automatically captured in the studio. The designer is responsible for assessing the fastener visibility. Flexible viewing controls have been built into the studio to aid in this task.

8.3.2 Disassembly process simulation in simultaneous product/process design

Emerging from VP generation is a virtual prototype that is suitable for simulation and assessment. Virtual prototype animations for disassembly simulations in simultaneous product/process design are presented in this section.

Disassembly process simulation capabilities

The generation of disassembly processes and the simulation of those processes are two different issues in this work. Currently, the ability exists to execute virtual prototypes consisting of sequential as well as simultaneous disassembly operations. Two descriptors of the disassembly process are needed for the simulation, namely a disassembly process sequence and the corresponding disassembly paths. In our system, disassembly paths are represented by parametric curves and manipulated through control vertices. Using the process model, a designer can simulate the process in two modes: interactive graphical or CNG (see no graphics).

In the interactive graphical environment, disassembly processes are graphically animated for inspection and manipulation. In addition, disassembly paths (along with their control vertices) can be displayed with the product models in the graphics window. More importantly, the system user can interact with the disassembly paths by clicking and dragging control vertices to new positions. The interactive graphical environment has proven to be an interesting design tool for rapid assessment of disassembly processes. In Fig. 8.8 a screen dump of a user manipulating a control vertex defining the endcap's disassembly path around an obstacle illustrates a simple disassembly simulation.

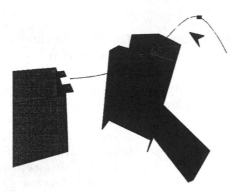

Fig. 8.8 Disassembly path manipulation.

The CNG mode is the second form of simulation in this work. CNG simulations are designed specifically for the evaluation of process constraints and goals during the solution of simultaneous product/ process design problems (see below). Whereas the graphical simulations involve a virtual prototype's graphics models, the CNG simulations utilize CODA's solid geometry product models. Using CSG models in the simulations is computationally more expensive, but it is required to support the necessary evaluation of geometric constraints and goals. For instance, a Boolean intersection is used to perform collision detection during simulation.

Representing paths with parametric curves

In this work, a disassembly path is defined as any path traversed from point A to point B. Moreover, disassembly paths can represent either a component's disassembly path or a path defined by a disassembler removing a fastener. To date, we have implemented two parametric curves formulations in our system, namely the cubic Bezier and spline curves. Due to space limitations, we will not delve into the mathematics behind these curves but refer the reader to Mortenson (1985) and Press *et al.* (1988). Although both curves have been implemented in our system, we prefer cubic splines because of the ability to select exact points through which the curve must pass.

The only drawback to this path representation is that a constant step size in parametric space seldom corresponds to a constant step size in Cartesian space. As a result, a constant step in the parametric variable u may result in an unacceptably large or undesirably small step sizes in Cartesian space. To minimize this characteristic, we constrain the average Cartesian translation and orientation step sizes to a user-defined maximum value based on each object's minimum geometric dimension.

8.3.3 Formulation of GDG-DSP for simultaneous product/process design

Integral to our approach to designing for disassembly via virtual prototypes is the development of parametric decision support problems in which individual product and disassembly process design problems are integrated and solved simultaneously with the aid of virtual prototype simulations and optimization techniques. The decision support problems in simultaneous product and disassembly process design (SPDP design) take the form of the goal-directed geometry decision support problem (gdg-DSP). Goal-directed geometry (GDG) is a framework for the development and solution of optimization problems involving geometry (Rosen *et al.*, 1994). The term goal-directed geometry refers to the formulation of parametric decision support problems coupled to a geometric modeler and their solution using an extension of goal programming. This framework has been applied in the parametric design stage, where the system configuration and parameters are known but the specific values of those parameters have yet to be determined. It is intended to aid a designer confronted with 'what if' questions, examining the changes to the solution if (for instance) new constraints are added, if old constraints are removed, or if parameter values are fixed. The goal-directed geometry decision support problem is a particularization of the compromise decision support problem (Mistree *et al.*, 1993) which refers to a class of constrained, multi-objective optimization problems, where values of the design variables are determined that satisfy a set of constraints and achieve as closely as possible a set of conflicting goals. The compromise DSP is stated in words as follows.

Given
    an alternative that is to be improved through modification
    assumptions used to model the domain of interest
    the system parameters
    the goals for the design

Find

the values of the independent system variables (they describe the attributes of an artifact)

the values of the deviation variables (they indicate the extent to which the goals are achieved)

Satisfy

the system constraints that must be satisfied for the solution to be feasible

the system goals that must achieve a specified target value to the extent possible

the upper and lower bounds on the system variables

Minimize

the deviation function Z which is a measure of the deviation of the system performance from that implied by the set of goals and their associated priority levels or relative weights.

A pictorial representation of the goal-directed geometry decision support problem as particularized for simultaneous product/process design is presented in Fig. 8.9. This particularization of the goal-directed

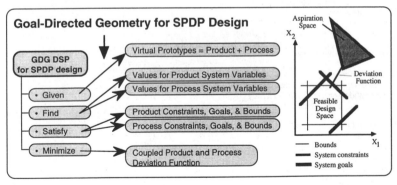

Fig. 8.9 Pictorial representation of GDG in SPDP design.

geometry decision support problem for simultaneous product/process design is further described below. Then the process constraints and goals which are used to evaluate disassembly processes in simultaneous product/process design are discussed.

Goal directed geometry for simultaneous product/process design

As stated earlier, goal-directed geometry is the decision support environment in which product design and disassembly process design are integrated, and trade-off assessments between product and process requirements are performed. The goal-directed geometry DSP for simultaneous product/process design (illustrated pictorially in Fig. 8.9 and mathematically in Fig. 8.10) is a particularization of the baseline gdg-DSP developed by Rosen *et al.* (1994) which is in turn a particularization of the compromise DSP. The particularizations which distinguish the gdg-DSP in simultaneous product/process design from that of the baseline gdg-DSP developed by Rosen and coworkers are as follows:

- inclusion of virtual prototypes in addition to geometric bodies
- definition of disassembly process design variables
- formulation of disassembly process constraints and goals
- definition of coupled product/process deviation function

Virtual prototypes are included in the gdg-DSPs in simultaneous product/process design to enable the evaluation of disassembly process constraints

---

Given
An alternative to be improved through modification.
Assumptions used to model the domain of interest.
A virtual prototype for disassembly:
    a parameterized, feature-based product model
    composed of geometric bodies, B_k, $k = 1, ... , t$
    a parameterized disassembly process model
System variables, X_j, $i = 1, ... , p$
Algebraic system constraint function, $g_i(\underline{X})$:
    $g_i(\underline{X}) = C_i(\underline{X}) - D_i(\underline{X})$
$f_k(d_i)$ function of deviation variables to be minimized at priority level k
    for the preemptive case.

Find
Product design variable values, X_j, $i = 1, ... , n$
Process design variable values, X_j, $i = n+1, ... , p$
Dependent design variable values, X_j, $i = p+1, ... , u$
Deviation variable values, d_i^-, d_i^+ $i = 1, ... , q$

Satisfy
Product and process algebraic constraints (linear, nonlinear)
    $g_i(\underline{X}) \geq 0$; $i = 1, ... , r$
Product and process geometric constraints (linear, nonlinear)
    $g_i(\underline{X}, \underline{B}) \geq 0$ $i = r+1, ... , s$
Product and process algebraic goals (linear, nonlinear)
    $g_i(\underline{X}) + d_i^- - d_i^+ = G_i$; $i = 1, ... , m$
Product and process geometric goals (linear, nonlinear)
    $g_i(\underline{X}, \underline{B}) + d_i^- - d_i^+ = G_i$; $i = m+1, ... , q$
Bounds
    $X_i^{min} \leq X_i \leq X_i^{max}$; $i = 1, ..., p$
    $d_i^-, d_i^+ \geq 0$; $i = 1, ..., q$
    $d_i^- \cdot d_i^+ = 0$; $i = 1, ..., q$

Minimize
Coupled product and process preemptive deviation function

    $Z = [f_1(d_i^-, d_i^+), ..., f_k(d_i^-, d_i^+)]$

Coupled product and process Archimedean deviation function

    $Z = \sum W_i(d_i^- + d_i^+)$ where $\sum W_i = 1, W_i \geq 0$

Fig. 8.10 Mathematical form of GDG-DSP in simultaneous product/process design.

and goals without the need for physical prototypes. A detailed description of virtual prototypes and virtual prototyping is presented here. Disassembly process system variables are included in the gdg-DSP in simultaneous product/process design to enable the modification of the disassembly process and the performance of trade-off assessments between product and process requirements. An example of the disassembly process system variables in this work is the control points which define a disassembly path. To evaluate the feasibility and goodness of a disassembly process, disassembly process constraints and goals are included in the gdg-DSP for simultaneous product/process design. Rather than develop physical prototypes, virtual prototypes are used in the evaluation of the disassembly process constraints and goals. The deviation from the target values for the disassembly process goals are combined with product deviation variables to form a coupled product/process deviation function which drive product/process improvement. With the aid of optimization tools, a solution is located which satisfies the product/process constraints and bounds and achieves to the extent possible the product/process goals as modeled in the deviation function. The deviation function can be implemented in two ways: the pre-emptive and Archimedean forms.

Disassembly process goals and constraints in GDG

In their most rudimentary form, disassembly process constraints and goals are formulated to force the disassembly process to a state of non-interference and minimum energy. In our current formulations, non-interference is the standard disassembly process constraint. Therefore, a disassembly process is deemed feasible if it results in product disassembly and avoids physical interference between objects in the simulation environment. In addition to the non-interference constraint, we utilize three disassembly process goals to force the disassembly process's energy to a minimum state: minimize the disassembly path lengths, minimize the work done over the path, and minimize the reorientation of component and tools. A fourth goal, which often conflicts with the previous goals is as follows: maximize the clearance between components and obstacles along their respective disassembly paths.

Overall, disassembly process constraints and goals are only limited by one's imagination and evaluation capability (i.e. simulation capability) and computational resources. For instance, if robotic disassembly is utilized, additional constraints would be added to govern the kinematics of the robot. The ability to develop, utilize and evaluate detailed simulations is constrained by one's available computational resources, as well as the current level of design knowledge. In the early stages of design, it is advisable to have simple process constraints/goals and simulations, because high product uncertainty and multiple conceptual designs can easily strain one's resources. As the design process progresses, design knowledge increases, and the number of design concepts decreases and more detailed simulations can be developed to enable the evaluation of more complex and meaningful process constraints and goals.

8.3.4 Solution algorithm for simultaneous product and disassembly process design

The ALP algorithm (Mistree *et al.*, 1993a) which is resident in DSIDES (decision support in design of engineering systems) (Reddy *et al.*, 1992) is the traditional solver of compromise DSPs in the Systems Realization Laboratory at Georgia Tech. Because the ALP algorithm has proven to be an ineffective solver of the goal-directed geometry problem in simultaneous product/process design, a multiobjective pattern search algorithm, based on the Hooke and Jeeves pattern search method (Reklaitis *et al.*, 1983), has been developed for the solution of this class of GDG problems. Unlike the ALP algorithm, which solves an approximate problem exactly, the Hooke–Jeeves (HJ) pattern search method is a method that solves the exact problem approximately. Basically, the HJ procedure is a combination of exploratory moves of the one-variable-at-a-time kind, with pattern or acceleration moves regulated by some heuristic rules (Reklaitis *et al.*, 1983). The exploratory moves examine the local behavior of the objective function and seek to locate the direction of any sloping valleys that might be present. The pattern moves utilize the information generated in the exploration to step rapidly along the valleys.

Illustrated in Fig. 8.11 is a flowchart of the implementation of the HJ pattern search algorithm in this work. As can be seen in Fig. 8.11, the

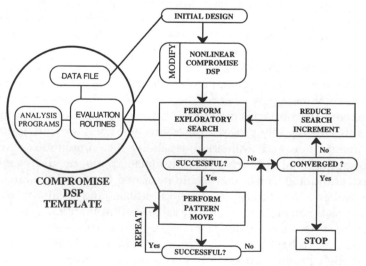

Fig. 8.11 Implementation of HJ pattern search for solving compromise DSPs.

compromise DSP template has been integrated with the traditional HJ pattern search method, and a lexicographical minimum replaces the traditional single objective formulation, thus allowing the modeling and solution of c-DSPs and gdg-DSPs. (A compromise DSP template is a mathematical model of a compromise DSP which is expressed in terms of variables, constraints, goals, etc., and is therefore implementable on a computer.)

8.4 AN ILLUSTRATIVE EXAMPLE

We illustrate our approach to design for disassembly via virtual prototyping through an example. Assume that we wish to develop an automotive center console, shown in Fig. 8.12, that is easily demanufactured into reusable, recyclable and/or scrapable modules. Several components (e.g. the bin, ashtray, hinge and coverplate) are definite candidates for reuse because they are hidden from view, and their

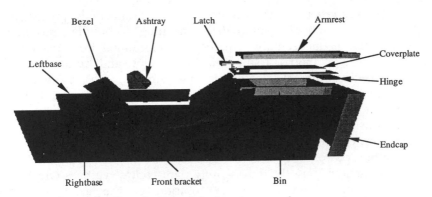

Fig. 8.12 Conceptual product model of center console.

appearance means very little in terms of customer satisfaction. The remaining components will probably be recycled because they are not compatible with next year's interior redesign. This problem was selected because it is a practical example of the types of system that are excellent candidates for demanufacture. In the remainder of this section, the steps of simultaneous product/process design as outlined in Fig. 8.2 are applied to the design of the center console's rear module and disassembly process.

8.4.1 Step 1 in SPDP design, definition of the problem

Given the conceptual design of the center console as shown in Fig. 8.12, it is our goal to integrate the design of the bin, endcap and armrest components (i.e. rear module of center console) with the design of their disassembly process. The rear module is illustrated in Fig. 8.6. In this scenario, the product design problem consists of determining optimal dimensions of the bin to maximize storage capacity and minimize the subassembly's material volume. Ensuring adequate wall thickness to support anticipated loading conditions and facilitate injection molding is the only product constraint explicitly defined in this design problem. All other product constraints are implemented as bounds on the system variables. Process goals include minimizing the arc length of travel necessary to open the armrest during disassembly and minimizing disassembly paths for the disassembler, in addition to the components being removed, while avoiding physical interference between objects during disassembly.

8.4.2 Step 2 in SPDP design, generation of virtual prototype simulations

In the second step of simultaneous product/process design, the virtual prototype simulations of the center console's disassembly process are generated. Through the automated reasoning techniques discussed above, a partial disassembly sequence is determined from the center console's product model. Using this information and human-assisted methods outlined above, a complete disassembly process is generated in the virtual environment, as outlined in Table 8.2. Also listed in Table 8.2 are the tools necessary for each operation, in addition to short descriptions of each operation. The information generated in the virtual prototyping environment is used to create real-time simulations of the disassembly process for use in solution of the GDG problem. Not shown in Table 8.2 are the disassembly path control vertices which are used as initial estimates in the solution of the example GDG problem.

8.4.3 Step 3 in SPDP design, formulation of goal-directed geometry DSP

In the third step of simultaneous product/process design, the goal-directed geometry DSP is formulated that mathematically represents the product (center console's rear module), the process (disassembly of rear module), design requirements (constraints and goals), as well as the designer's preferences (merit function). The gdg-DSP for this problem is presented in Fig. 8.13. Now that the necessary virtual prototype simulations and mathematical form of the gdg-DSP have been developed, the simultaneous product/process design problem is implemented on a computer and solved.

Table 8.2 Center console's rear module's disassembly process

Task no.	Part name	Description of task	Direction of removal and unfastening	Tool	Comments
1	Armrest	Open armrest	[0, 0, 0]	None	Rotate armrest 120 deg about y axis
2	Screw 1	Unfasten, remove	[0, 0, 1] [0, 0, 1]	Screwdriver	Rotate screwdriver about +z axis and remove fastener in +z direction
3	Screw 2	Unfasten, remove	[0, 0, 1] [0, 0, 1]	Screwdriver	Rotate screwdriver about +z axis and remove fastener in +z direction
4	Armrest	Remove	[0, 0, 1]	None	Remove armrest in the +z direction
5	Endcap	Remove	[-1, 0, 0]	None	Remove endcap in the -x direction

Given:
  CAD model of center console's rear module: Disassembly process model for center console's rear module:
    rear module component's include endcap,   geometric model of disassembler's hand grasping
    armrest, and bin.   screwdriver.
    parameterized geometric models of rear module.   process disassembly operations and sequences for rear
    dimensional relationships between components.   module.
    mating relationships between components.   parameterized disassembly paths for rear module.
  Flow path ratio: ratio of length of flow per unit wall thickness

Find:
  Values for dimensions defining bin's   Values for disassembly process parameters:
    overall geometric shape: endcap's control vertices $(\alpha_1\, \rho_1\, \gamma_1\, x_1\, y_1\, z_1)$ $(\alpha_2\, \rho_2\, \gamma_2\, x_2\, y_2\, z_2)$
    bin's length (L).   armrest's disassembly path control vertices $(\alpha_3\, \rho_3\, \gamma_3\, x_3\, y_3\, z_3)$
    bin's width (W) disassembler's path control vertices $(\alpha_4\, \rho_4\, \gamma_4\, x_4\, y_4\, z_4)$
    bin's height (H)
    bin's wall thickness (WT)
    armrest's unfastening angle $(\Delta\alpha)$

Satisfy:
  <u>Bounds:</u>
    $24.0 \leq L \leq 30.0$ cm $10.5 \leq W \leq 16.5$ cm $13.5 \leq H \leq 19.5$ cm $-2\pi \leq \Delta\alpha \leq 2\pi$ rad.
    $-2\pi \leq (\alpha_i,\, \rho_i,\, \gamma_i) \leq 2\pi$ rad   for $i = 1..4$ $-200 \leq (x_i,\, y_i,\, z_i) \leq 200$ cm   for $i = 1..4$
  <u>Constraints:</u>
  • Non-interference constraint

$$\sum_{i=1}^{N_O}\left[\sum_{j=0}^{N_s^i} \frac{volume(comp_{i,j} \cap environment)}{volume(comp_{i,j})}\right] = 0$$

  where N_O is the number of disassembly operations,
  N_s^i is the number of steps in the ith disassembly operation,
  $comp_{i,j}$ is the object associated with the ith operation on the jth step,
  and *environment* is all objects not associated with $comp_{i,j}$

  <u>Goals:</u>
  • Bin's storage volume goal (maximize)   • Assembly's material volume goal (minimize)

$$\frac{storage\_volume}{target_{storage\_volume}} - 1 = d_1^+ - d_1^-$$
$$1 - \sum_{i=0}^{num\ comps}\left(\frac{target_{volume(comp_i)}}{volume(comp_i)}\right) = d_2^+ - d_2^-$$

  • Armrest's arc length goal (minimize)   • Disassembly path length goal (minimize)

$$1 - \frac{target_{arc\_length}}{arc\_length} = d_6^+ - d_6^-$$
$$\left[1 - \frac{\left|\mathbf{P}_{N_s^i}^i - \mathbf{P}_0^i\right|}{\sum_{j=1}^{N_s^i}\left|\mathbf{P}_j^i - \mathbf{P}_{j-1}^i\right|}\right]_{i=1..N_O} = d_{i+2}^+ - d_{i+2}^-$$

  • Endcap's obstacle clearance distance goal (achieve)

$$\frac{\min\limits_{i=1}^{3}\left(\left|\mathbf{W}_j^i - \mathbf{R}\right|\right)_{j=1..N_s^i}}{target_{clearance\_dist}} - 1 = d_7^+ - d_7^-$$

  where $\mathbf{W}_j^i$ is the endcap's i^{th} point on j^i
  $\mathbf{R}$ is a point of reference on an ol
  for $i = 1..N_O$; where $\mathbf{P}_j^i$ is j^{th} position in i^{th} operation

Minimize:

$$Z = \left\{\begin{matrix}(0.6 \cdot d_1^- + 0.4 \cdot d_2^+)/2 \\ (d_3^+ + d_4^+ + d_5^+ + d_6^+)/4 \\ (d_7^-)\end{matrix}\right\}$$

NOTE: The variable, d, represents the distance (deviation) between the aspiration level and the actual attainment of a goal. The deviation variable, d, can be positive or negative depending on whether over or under achievement has occurred Therefore, the deviation variable is replaced with two variables: $d = d_i^- - d_i^+$ where $d_i^- \cdot d_i^+ = 0$ and $d_i^-, d_i^+ \geq 0$. Accordingly, the objective in the C-DSP is to minimize the deviation from target.

Fig. 8.13 GDG problem for simultaneous product/process design of rear module.

8.4.4 Step 4 in SPDP design solution of SPDP design problem

To enable simultaneous product/process design, parametric disassembly simulations are synthesized from the disassembly process information gathered in the virtual prototyping environment. As outlined above, these simulations are essential to the solution of the GDG problem. By using the solution process described above and by assuming an injection molding flow path ratio of 115 (Cracknell and Dyson, 1993), 'optimized' solutions are located for the simultaneous product/process design problem presented in Fig. 8.13. Tables 8.3 and 8.4 contain the initial and final values for the design variables. Table 8.3 pertains to product design variables, and Table 8.4 contains process design variables. The reason why several values for the process design variables remain unchanged from their initial values is because the variables' final values were obvious to us during problem formulation and thus were set to the final values to minimize the computational expense involved in solving an optimization problem involving 28 design variables, 3 constraints, 8 goals, and computationally expensive simulations.

Table 8.3 Product design variables: initial and final values

Design var.	Initial value	Final value
L (cm)	27.0	27.3
W (cm)	13.5	15.1
H (cm)	16.5	18.45
WT (cm)	0.27	0.30
$\Delta\alpha$ (rad)	2.69	1.52

The impact of our problem formulation and solution scheme on the design of the center console's rear module is described in Table 8.5, which includes the initial and final values for key design drivers and percentage improvements over their initial values. As shown, significant improvements were realized with respect to all but one of the key design drivers. The bin's interior volume goal dominated the assembly's material volume goals because of the weighting scenario in the first priority level of the deviation function (refer to Fig. 8.13). The convergence of the three priority levels of the deviation function is shown in Fig. 8.14, where the level number corresponds to deviation function levels (i.e. merit function levels).

The proper selection of constraints and goals and the development of an accurate and representative deviation function is the key to our success in solving the GDG problem effectively. If the goals, constraints and deviation functions do not represent the designers' requirements and preferences, any solution obtained from the GDG problem will be meaningless. Therefore, we are of the opinion that multiple problem

Table 8.4 Disassembly process design variables: initial and final values

Design var.	Initial value	Final value	Design var.	Initial value	Final value	Design var.	Initial value	Final value	Design var.	Initial value	Final value
α_1 (cm.)	0.000	2.250	α_2	0.000	1.600	α_3	1.000	0.750	α_4	0.000	0.000
ρ_1 (cm.)	0.000	0.000	ρ_2	0.000	0.000	ρ_3	0.000	0.000	ρ_4	0.000	0.000
γ_1 (cm.)	0.000	0.000	γ_2	0.000	0.000	y_3	0.000	0.000	γ_4	0.000	0.000
x_1 (rad.)	−58.00	−60.00	x_2	−37.00	−41.00	x_3	18.25	18.25	x_4	15.00	3.00
y_1 (rad.)	20.00	3.00	y_2	20.00	9.13	y_3	−20.00	−10.75	y_4	0.00	0.00
z_1 (rad.)	40.00	35.00	z_2	45.00	40.40	z_3	30.00	20.00	z_4	35.00	22.97

Table 8.5 Product and disassembly process drivers: initial and final values

Design goals	Initial value	Final value	Percent improvement
Bin's interior volume (cm^3)	5558.0	7034.2	26.6%
Assembly's mat. volume (cm^3)	817.0	986.6	−20.8%
Armrest's arc length (cm)	54.0	41.50	23.1%
Endcap's path length (cm)	103.1	85.52	17.1%
Armrest's path length (cm)	57.6	40.04	30.5%
Disassembler's path length (cm)	73.2	56.37	23.0%

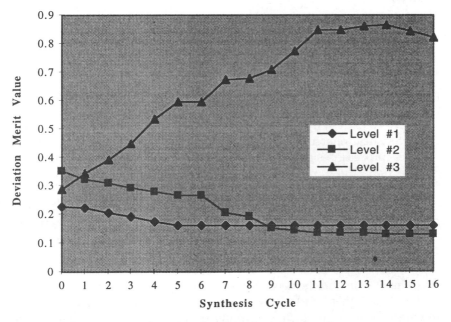

Fig. 8.14 Convergence of product/process deviation function.

formulations are warranted to determine a true solution, or better yet, a region of solutions. Consequently, the most difficult aspect of simultaneous product/process design via virtual prototyping is not the development of the virtual prototypes or simulations but the development of a GDG problem that is representative of our design requirements and preferences.

8.5 CHALLENGES IN SOLVING SIMULTANEOUS PRODUCT/PROCESS PROBLEMS

Although the solution process for simultaneous product and process design, as outlined above, is uncomplicated and intuitive, the application of optimization techniques to the solution of the supporting GDG problems has not been as transparent as one would hope. As stated above, two optimization algorithms have been implemented in this work for solution of goal-directed geometry problems: the ALP algorithm in DSIDES and a modified pattern search. Now several characteristics of the basic simultaneous product/process GDG problem are presented which have been problematic when applying these optimization techniques to their solution. Quick-fix solutions to these problems have been implemented, and are discussed below. Future plans for increasing the efficiency and effectiveness of the solution process are then briefly outlined.

8.5.1 Problematic characteristics of product/process problems in optimization

To increase the efficiency and effectiveness of simultaneous product and process design, optimization techniques are utilized in the solution of the supporting GDG problems. Ideally, optimization techniques are a plug-and-play type design tool, but unfortunately this is rarely true. Several characteristics inherent to the GDG problems supporting simultaneous product/process design have proven problematic. In the remainder of this section, these problems are brought to light, and subsequent solutions to these problems are presented.

The first and most problematic characteristic of the simultaneous product/process GDG problem is the necessity to couple design variables to successfully evaluate trade-offs among design goals. Unfortunately, the ALP and pattern search optimization algorithms do not provide this capability, and are only effective problem solvers when design variables are independent and the necessity for coupling among design variables is negligible. In other words, these optimization techniques are incapable of solving problems in which a coupling of design variables is required to properly evaluate the objective function. In this class of GDG problems, the most apparent illustration of this fact is the coupling that must occur between design variables that define a component's disassembly path and those that define its orientation along that path. To properly assess trade-offs among design goals (such as minimizing path length, potential energy and degree of rotation), the coupling of translational and rotational design variables is necessary, as illustrated in Fig. 8.15. In this simple scenario, an object traverses a path from its initial position to its final position while avoiding an obstacle. If the disassembly process

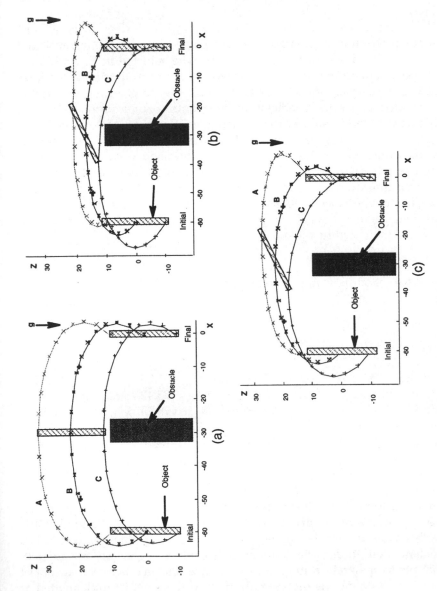

Fig. 8.15 Interactions among process design variables.

constraints, goals and optimization techniques, such as the ALP or pattern search methods, are applied to this problem, an optimum solution would be determined in which the object remains in its initial upright orientation and just avoids a collision with the obstacle, as shown in Fig. 8.15(*a*).

A more intuitive solution is illustrated in Fig. 8.15(*b*) in which the object's position and orientation have changed, resulting in a shortened path length and less work over the path. The solution in Fig. 8.15(*b*) is difficult to obtain in the ALP or pattern search optimization techniques because the impact of each design variable on the objective function is determined in a one-variable-at-a-time manner. For example, a design variable that controls the object's orientation along the path is perturbed in each direction (+/− a delta rotation) as shown in Fig. 8.15(*c*), and its impact on the objective function is observed to increase the average path length and degree of rotation. As no coupling of design variables occurs, the algorithm does not realize that a change in orientation may enable a lowering of the object's center of gravity and subsequent decreases in path length and work. The implications of the scenarios on the process goals are tabulated in Table 8.6.

Table 8.6 Comparison path selection on goals

Process design consideration	Fig. 8.15(a)	Fig. 8.15(b)	Fig. 8.15(c)
Path length A	97.38	81.33	93.55
Path length B	97.38	85.42	97.40
Path length C	97.38	95.23	106.81
Avg. path length	97.38	87.33	99.25
Max height @ C.G.	20.00	18.00	20.00

In order to facilitate the location of a solution such as presented in Fig. 8.15(*b*), a minimum clearance goal (Martinez-Alfaro and Flugrad, 1994) has been added to the general simultaneous product/process GDG problem definition. The minimum clearance goal, as formulated, attempts to maintain a minimum clearance distance between an object and obstacles along its disassembly path. If this goal is implemented at the same priority level as the minimum path length goal in a pre-emptive objective formulation, a trade-off, in theory, would occur between the minimum path length goal and the minimum clearance distance goal. However, if the minimum clearance distance goal is implemented at a lower priority level than the minimum path length goal in a pre-emptive formulation, the minimum clearance goal will not interfere with the minimum path length goal (assuming a single point model for path length calculations). Instead, the minimum clearance goal will force the object to rotate away from the obstacle, thus enabling the object to get closer to the obstacle, as shown in Fig. 8.15(*b*). In the example problem illustrated above, the third level objective function consists of the d^- deviation variable for the minimum clearance goal. As can be seen in Fig. 8.14, the value of the third priority level deviation function increases throughout the optimization process, whereas the values of the first and second priority levels converge sequentially to their final values. Although the minimum clearance goal enables the manipulation of orientation to decrease the object's path length, it does not enable the assessment of trade-offs between path length, work, and degree of rotation because changes in orientation are still independent of translation, and vice versa. To perform a true trade-off between the process goals, a coupling of process design variables needs to occur.

A second characteristic that has proven problematic to the solution process is an interaction between product model geometry and the associated disassembly. What we have discovered is that the product model will unnecessarily constrain the disassembly paths, and vice versa, if one set of design variables converges on their solution faster than another set. For example, consider the center console's rear module presented above. As the armrest's disassembly path becomes shorter and closer to the bin, it will unnecessarily constrain the height of the bin because an increase in bin height will cause interference during the disassembly path. As one may deduce, this also is a problem resulting from the inability to couple design variables during the solution process. To minimize this characteristic, three obvious steps can be taken. First, the optimization's step sizes can be adjusted to facilitate a quicker convergence for the product model. A better fix is the addition of a minimum clearance goal at the same level as the minimum path length goal. The addition of minimum clearance goal will provide a region of flexibility (or buffer zone) to the solution process, minimizing undesirable physical interactions between product and process models. Unfortunately, our current solid geometric modeler does not support a minimum distance calculation and as a result, implementation of the minimum clearance goal entails additional difficulties. The third option, which is possibly the best, is the development of an optimization algorithm that can solve design problems which require a coupling of design variables to be solved correctly.

A third problematic characteristic of the simultaneous product/process GDG problems is related to the ON/OFF behavior associated with the non-interference constraint. If a non-interference constraint is satisfied, the value of the constraint is zero. On the other hand, when interference occurs between two objects, the non-interference constraint is proportional to the volume of interference. As a result, our non-interference constraint is piecewise continuous, and its derivative is discontinuous as two objects contact one another. This poses a significant problem for the linearization scheme utilized by the ALP algorithm resident in DSIDES (Mistree *et al.*, 1992). The ALP algorithm perturbs each design variable by an incremental amount to determine its relationships with the constraints and goals of the design problem. If the perturbations do not result in interference, the ALP algorithm assumes that the non-interference constraint is independent of those variables and suppresses the non-interference constraint while it solves a linearized approximation of the design space. The solution of the linearized problem often resides in a region of high geometric interference.

A fourth characteristic of this class of GDG problems which has caused problems for DSIDES's ALP algorithm is the relatively large and underconstrained design space. Why is the design space so large and underconstrained? The design space is large because of the number of design variables required to control an object's position and orientation along a disassembly path is large. For instance, to control a path in three dimensions with a third-order parametric equation requires 12 design variables. The ranges on the process design variables are necessarily large, which further hinders DSIDES linearization. DSIDES's ALP algorithm is most efficient and effective at solving problems characterized as having highly constrained and small design spaces.

Due to the ON/OFF behavior of the non-interference constraint and the size and shape of our design spaces, DSIDES's ALP has been unable to solve any simultaneous product/process design problems. The multiobjective pattern search algorithm described above has been more successful than the ALP algorithm because it is not hindered by the difficulties associated with ALP's linearization scheme. Nonetheless, the pattern search may prove ineffective and inefficient as the size of the decision support problem increases and the need/desire to properly assess trade-offs among goals is heightened. Our plans for dealing with these difficulties are highlighted in the next section.

8.5.2 Future plans

Two issues need to be further addressed if the efficiency and effectiveness of the solution process is to increase. The first is the coupling of design variables which must occur if proper trade-off among design goals are to occur. To help resolve this problem, a switch from conventional optimization to non-conventional optimization techniques such as genetic algorithms or simulated annealing may be beneficial. (A multiobjective simulated annealing algorithm with downhill-simplex is being implemented in this work. Initial results indicate that the algorithm is effective but not necessarily efficient at solving simultaneous product/ process design problems). The second issue that needs to be addressed is the inability to accurately calculate the minimum clearance goal in the GDG problem formulation. As stated previously, our current solid modeler does not support this calculation. To improve the efficiency of our solution process, an upgrade to the NURBS-based solid modeler ACIS is in progress. ACIS does support a minimum clearance calculation. Additional improvements and modifications to our solution process will be realized as we continue to develop a better understanding of the class of multiobjective optimization problems that we are encountering in simultaneous product and process design.

8.6 CONCLUSION

Within the context of product design for the life cycle, disassembly plays an important role in product service, material recycling and component reuse. To facilitate rapid product development in light of these life cycle considerations, we propose to design disassemblability into products through the use of simultaneous product/process design problems via virtual prototyping. As outlined in this work, we have developed a class of problems that includes both product and process design considerations. By first ensuring the satisfaction of product functionality, disassembly processes can be improved by utilizing remaining design freedoms to identify product parameter values that are favorable to disassembly. We have described our approach to the development of disassembly processes through the use of virtual prototyping techniques and presented the product/process decision support problems in this chapter. We are currently applying this methodology to the design of an automotive center console. The results to date indicate that virtual prototyping greatly facilitates the development of disassembly processes for use in evaluation of the simultaneous product/process decision support problem. Preliminary results indicate that the multiobjective solution method does yield improved product and process designs; however, the solution of these problems is sensitive to the presence of complex geometric obstacles to disassembly and the necessity to couple design variables to properly assess trade-offs between product and process requirements.

The approach presented in this chapter is limited to parametric design for disassembly. Future work will include fastener selection as part of the product/process problem formulation, because fasteners strongly influence the efficiency and effectiveness of disassembly processes. Fortunately, the development of a coupled selection and compromise decision support problem will enable fastener considerations, as well as parametric synthesis of products and their disassembly simulations. An additional subject for future work is the development of a robust solution mechanism for goal-directed geometry decision support problems of the class developed in simultaneous product/process design.

ACKNOWLEDGMENTS

We acknowledge the support of the National Science Foundation through grant DMI–9420405 and DMI–9414715.

REFERENCES

Beitz, W. (1993) Designing for ease of recycling – general approach and industrial application. *Ninth International Conference on Engineering Design*, The Hague, Netherlands, HEURISTA. **2**(3), 731–738.

Boothroyd, G., and Dewhurst, P. (1991) *Product Design for Assembly*, Wakefield, Boothroyd and Dewhurst, Inc.

Chen, L.-L., and Woo, T. C. (1992) Computational geometry on the sphere with application to automated machining. *Journal of Mechanical Design*, **114**, 288–295.

Connacher, H. I., Jayaram, S., and Lyons, K. (1995) Virtual assembly design environment. *Proceedings of the Computers in Engineering Conference and the Engineering Database Symposium*, Boston, MA, 875–885.

Coulter, S., Bras, B., and Rosen, D. (1995) Dynamic non-interference constraints in goal-directed geometry. *Advances in Design Automation, 21st Design Automation Conference*, Boston, Massachusetts, 907–914.

Cracknell, P. S., and Dyson, R. W. (1993) *Handbook of Thermoplastics Injection Mould Design*, Blackie Academic & Professionals, New York.

Dai, F., and Gobel, M. (1994) Virtual prototyping – an approach using VR-techniques. *Proceedings of the Computers in Engineering Conference*, Minneapolis, 311–316.

De Mello, L. S. H., and Sanderson, A. C. (1988) Planning repair sequences using the AND/OR graph representation of assembly plans. *Proceedings of 1988 IEEE International Conference on Robotics and Automation*, 1588–1593.

Donald, B. R. (1987) A search algorithm for motion planning with six degrees of freedom. *Artificial Intelligence Journal*, **31**, 295–353.

Hoffman, R. (1990) Assembly planning for B-rep objects. *Proceedings of the Rensselaer 2nd International Conference on Computer Integrated Manufacturing*, 314–321.

Huang, K. I. (1993) Development of an assembly planner using decomposition approach. *Proceedings of the IEEE International Conference on Robotics and Automation*, 527–532.

Keirouz, W., Pabon, J., and Young, R. (1990) Integrating parametric geometry, features, and variational modeling for conceptual design. *ASME Design Theory and Methodology Conference*, Chicago, 1–9.

Kroll, E., Lenz, E., and Wolberg, J. R. (1988) A knowledge-based solution to the design for assembly problem. *Manufacturing Review*, **1**(2), 104–108.

Kuehne, R. P., and Oliver, J. H. (1995) A virtual environment for interactive assembly planning and evaluation. *Design Engineering Technical Conferences*, Boston, Massachusetts, 863–867.

Lee, S., and Shin Y. G. (1990) Assembly planning based on subassembly extraction. *Proceedings of the IEEE International Conference on Robotics and Automation*, 306–313.

Light, R., and Gossard, D. (1982) Modification of geometric models through variational geometry. *Computer-Aided Design*, 14, 209–214.

Marks, M., Ishii, K., and Eubanks, C. F. (1993) Evaluation methodology for post-manufacturing issues in life-cycle design. *Concurrent Engineering: Research and Applications*, 1(1), 61–68.

Martinez-Alfaro, H., and Flugrad, D. R. (1994) Collision-free path planning for mobile robots and/or AGVs using simulated annealing. *IEEE*, 270–275.

Mistree, F., Smith, W. F. Bras, B., Allen, J. K., and Muster, D. (1990) Decision-based design: a contemporary paradigm for ship design. *Transactions on the Society of Naval Architects and Marine Engineers*, 565–597.

Mistree, F., Hughes, O. F., and Bras, B. A. (1992) The compromise decision support problem and the adaptive linear programming algorithm. *Structural Optimization: Status and Promise*, AIAA, Washington, D. C., 251–290.

Mistree, F., Smith, W. F., and Bras, B. A. (1993) A decision-based approach to concurrent engineering. *Handbook of Concurrent Engineering*, Chapman & Hall, New York, 127–158.

Mortenson, M. E. (1985) *Geometric Modeling*, John Wiley, New York.

Press, W. H., Flannery, B. P., Teukolsky, S. A., and Vetterling, W. T. (1988) *Numerical Recipes in C*, Cambridge University Press, New York.

Reklaitis, G. V., Ravindran, A., and Ragsdell, K. M. (1983) *Engineering Optimization: Methods and Applications*, John Wiley, New York.

Rosen, D. W., Chen, W., Coulter, S., and Vadde, S. (1994) Goal-directed geometry: beyond parametric and variational geometry CAD technologies. *ASME Design Automation Conference*, 417–426.

Woo, T. C., and Dutta, D. (1991) Automatic disassembly and total ordering in three dimensions. *Journal of Engineering for Industry*, 113, 207–213.

Zussman, E., Shoham, M., and Lenz, E. (1992) A kinematic approach to automatic assembly planning analysis. *Manufacturing Review*, 5(4), 293–304.

CHAPTER NINE

Human aspects of rapid response manufacturing

John V. Draper

9.1 INTRODUCTION

In the first decades of the 19th century, British industry was experiencing a technological revolution. The rise of the factory system meant that groups of people were now concentrating at a single location to operate the new machines that were at the heart of the production system. The old system, in which skilled workers who were largely self-employed produced goods by their hands within their own homes, was rapidly waning. The social displacement caused by the imposition of the new system led to the Luddite riots, in which angry craftsmen destroyed what machinery they were able to lay their hands on in the industrial midlands. They did this as an expression of the anxiety provoked by the changes in social organization and because of the widespread belief that the new system contributed to unemployment.

In the last decades of the 20th century western industry is also experiencing a technological revolution. Machines are becoming more intelligent, with widespread use of computer control, and therefore technology is available to make production lines more flexible. This has led to the increasing application of what are called advanced manufacturing systems (AMS) or rapid response manufacturing. Technological change affected worker roles and the social structure that supported manufacturing two centuries ago, and in the future technological change will affect the jobs and social structures that exist in organizations today. It is yet to be seen whether the current revolution will have the impact of the earlier one, and it seems unlikely that it will provoke a similar destructive reaction among workers. However, it is certain that the introduction of advanced manufacturing practices will have an impact on the social organization and job structure of modern workers. Organizational structures developed to make best use of old-style production technology will not serve the new equipment and jobs designed to serve the old technology will change to meet the needs of the new machines, whether this change is planned or not.

New and emerging technologies provide an opportunity for future manufacturing systems to be much more flexible than existing systems in all phases of the design and production cycles. However, one component of these systems will not change: their human users. No treatment of rapid response manufacturing is complete without discussing the impact of the new technologies on the organization that provides the context for the machines, and on the demands that the organization and the machines place on human users.

The purpose of this chapter is to outline some of the sociotechnical and ergonomic implications of advanced manufacturing technology. The discussion is necessarily brief, given the limitations of space and the scope of the sociotechnical and ergonomic implications of advanced manufacturing systems. For a deeper treatment of these issues the interested reader should consult Karwowski *et al.* (1994), Salvendy and Karwowski (1994) and Gerwin and Kolodny (1992). This chapter concentrates on the sociotechnical aspects of AMS, with the goal of introducing the reader to key issues in organizing enterprises to best use AMS technology.

9.1.1 Human issues

Human factors engineering has a traditional emphasis on design, including function allocation and the design of hardware, software and workplace environments. However, in a broader sense human factors also apply to the design of organizations and jobs, and to personnel selection and training (Kidd and Corbett, 1988). Meshkati (1990) points out that optimal control of a complex human-machine system requires an integrated approach to design at three key levels: workstation, job and organization. The workstation is the total human–machine interface for a particular system, including displays, controls, seating, HVAC, etc. The job is the set of tasks that a particular human operator must perform. Organization is the pattern of relationships between workers. Technological systems that do not integrate these three aspects of the sociotechnical system will be prone to inefficiencies and malperformamance. Organizational development, selection and training are sometimes considered the purview of industrial/organizational psychology, but in the case of AMS the potential impact of the new technology on work organization has important ramifications for the traditional areas of human factors. Therefore, it is important to understand (and, hopefully, control) organizational dynamics to appropriately sculpt human–machine interfaces, allocate functions, etc. Furthermore, in a sense organizations are as much tools as hammers, computers or robots. Implementers of AMS need to create 'ergonomic organizations' as much as they need to create ergonomic workstations. It is also beneficial for traditional ergonomicists to understand the social milieu in which their more traditional activities take place to optimize their impact within an organization (Shipley, 1990).

Hendrick (1991, 1993) terms the traditional human factors activities 'microergonomics', and contends that they should be subsumed within a systems perspective that also considers organization and work design; organization and work design may be termed 'macroergonomics'. As maladjustments in either the microergonomic or macroergonomic system can affect mission success, planners and implementers of AMS should consider both.

Manufacturing enterprises are special social systems that either arise around or are designed to facilitate productive activities. Those who guide these enterprises have two tools with which they may control the production process: the social structure (formal and informal), which determines how information flows within the organization; and the equipment (hardware and software) in the hands of the members of the organization. In all cases, the equipment and the social structure mutually affect each other; both affect and are affected by their human users. A third resource for controlling the means of production is the human workers. They are amenable to change through training, education and motivation, but the boundaries within which people may change are more rigid than those within which equipment and social structures may change. Therefore, one may consider there to be a macroergonomic problem inherent in production and a microergonomic one. The macroergonomic problem has to do with designing the organization to best manage information and implement decisions. The microergonomic problem has to do with designing equipment to best fit the needs of the organization and human users. There is also a human resources issue, to do with providing adequate human skills through personnel selection and training. Solving the macroergonomic problem requires the development of institutional mechanisms for appropriately distributing information and acting on it. Solving the microergonomic problem requires the design of work environments and human–machine interfaces that promote safe and efficient use of the organization's equipment.

9.2 MACROERGONOMICS

Effective management of rapid response manufacturing technology requires more than replacing existing equipment on the shop floor with high-technology alternatives. It requires a basic restructuring of the manner in which the company controls its manufacturing process. The organization, its people and its equipment must be integrated into an efficient tool to control production. According to Gerwin and Leung (1986), the degree to which the organization supports the manufacturing technology 'may have a significant impact upon the ultimate success' of the AMS; and therefore 'managing this fit should be given priority.' A recent study of concurrent engineering supports this perspective; the devotion of management to planning organizational structure, tasks and implementation strategy is strongly correlated with organizational effectiveness (Duffy, Danek and Salvendy, 1995).

9.2.1 Structure

More has been written about how to implement AMS than about how to structure an organization to optimize AMS productivity. However, key issues in developing an advanced manufacturing organization to complement AMS technology may be enumerated. Hendrick (1987) identifies three components of the sociotechnical system as major influences on organizational structure. These are technology, environment and the personnel subsystem. Technology is the set of tasks to be performed, where a task may be defined as the mode of production, the actions that individuals perform to produce, or the strategy selected to reduce uncertainty. The environment is the economic and social milieu in which the organization functions. The personnel subsystem is the manner in which tasks are performed, including skills and training requirements, demographic characteristics of workers and organizational culture. According to Hendrick, technology is relatively fixed once selected, whereas the personnel subsystem must be flexible to allow the organization to adapt to the dynamic environment.

Pilitsis and Coringrato (1992) identify four influences on organizational structure: strategy, technology, people and environment. Strategy has to do with the organization's responses to events in the environment. Structure must be developed to facilitate achievement of strategic goals; structure also limits the strategies that an organization may adopt. Technology has to do with the means of production. Different technologies are best served by different organizations. For example, the flexibility inherent in AMS requires an adaptable organization that can deal with product and process changes with a minimum of disruption. Highly hierarchical structures impede adaptability. People, of course, are the members of the organization. They form a pool of human resources that must be taken into account in determining the appropriate structure. Structures cannot be successful if they demand more of members than they can contribute. The environment is the competitive context of the organization, and the degree of stability and predictability inherent in the environment are particularly important. Stable environments are less demanding; dynamic environments require the ability to adapt quickly.

Sun and Riis (1994), borrowing heavily from Leavitt (1972) and Child (1977), developed a model linking the organization, technology and management in advanced manufacturing technology implementation. The Sun and Riis model has four conceptual foundations:

1. organization, technology and strategy are interrelated
2. organizational and technological changes should be implemented synchronously
3. organization and technology should support the strategy
4. changes and interrelationships between the three factors need the intervention of management

These perspectives share some common ideas. First, the organization is the way that technology and personnel are integrated to form the means of production. Neither the technology nor the personnel subsystem is valuable in itself. It is only through the combination of the two that anything may be produced. Second, the various components of an organization are integrated into a complex system. No part of the organization is immune to changes occurring in another part, or to changes in the environment. Third, the organization must support the purposes of the enterprise. It must be capable of controlling production in a fashion that allows timely and efficient responses to environmental pressures. Fourth, the structure must be adapted to meet the capabilities of the personnel within the organization or available to it. Finally, the organizational structure is amenable to change by management and, more importantly, requires intervention by management to effectively meet the needs of the enterprise.

The consensus among researchers in this area seems to be that organizations must change to achieve optimal control over AMS and that it is important to achieve the appropriate balance between technology and personnel subsystems. Some recent research supports this contention (Horte and Lindberg, 1994). However, Hirsch-Kreinsen and Schultz-Wild (1990) in an examination of technological innovation in German firms found that 'interests and aims of management are not very clear-cut, if not even contradictory.' Managers in their study seemed to function with conflicting motives: to treat workers as potential sources of disturbances to be designed out of production wherever possible; and to depend on the availability of skilled workers to implement advanced technology and deal with shop floor disturbances. They concluded that five criteria are necessary for resolving these conflicting motivations and developing appropriate sociotechnical and ergonomic structures. These are that:

1. division of labor should be minimized
2. integrated approaches to job design must be developed
3. appropriate human–machine interfaces for each job must be developed, considering worker skills

4. training must be systematic and comprehensive
5. wage structures must be developed to prevent interference with team building and collective productivity

Organizations that emphasize distributed decision-making and problem-solving may be more suitable for dealing with the new dynamic means of production (Mathews, 1995; Engstrom *et al.*, 1995). Berniker (1990) points out that AMS depend more on highly skilled workers than conventional manufacturing approaches and on greater commitment from workers and management (and vendors) to the process. This is because AMS are not systems operated to a steady state, but rather are characterized by constant change. They are by design dynamic and the complexity, and strong interrelationships among components within the process increase the likelihood and impact of disturbances. These factors compel organizations to rethink the structure relating workers and management to each other. Structures should emphasize distributed authority and greater communication among workers at lower levels than may be the norm. According to Gerwin and Leung (1986), 'a semi-autonomous workgroup is an appropriate structure' for AMS and within this structure either job specialization or job rotation is appropriate, although each carries with it particular advantages and disadvantages. However, in a comparative survey of German and American firms, Harvey and Behr (1994) argue that although in Germany there is a tendency towards group work and away from job specialization, within American firms there is a greater tendency towards specialization. They attribute this to the traditional American approach of specialization and job reduction. This approach does not seem to foster the flexibility necessary for achieving optimal performance for an AMS. However, it is worth noting that in one study comparing two organizations that consciously attempted to balance technology, personnel and structure, to a third organization that maintained a traditional technology-centered approach, there were no apparent performance differences (Berger, 1994).

9.2.2 Tasks

Tasks may be thought of in a general sense to refer to the roles of members of the organization. In a very thorough examination of the impact of AMS on tasks, Robertson and Majchzark (1987) describe a set of first-order effects (directly affected by AMS implementation) and second-order effects (dependent on first-order decisions) of AMS technology on the personnel subsystem of a manufacturing organization. First-order effects occur in three areas, depending on the affected part of the work force: machine operators, technical support and management. The first-order effects for machine operators and technical support personnel are coordination needs, information needs, human-machine redundancy and autonomy. Simply put, AMS technology requires greater coordination and information sharing because of the close integration of production subsystems. This also requires more checking on machine operation by operators and greater operator autonomy in avoiding and responding to disturbances in the production process.

For managers and supervisors, first-order effects modify six functions: supervising (production technology and personnel), investigating, coordinating, planning and scheduling, staffing and training, and evaluating. The needs of the AMS and the first-order effects on workers jointly cause these effects. Managers and supervisors must become acclimatized to working with a technical and personnel system that requires supervision of a different type than is necessary on conventional shop floors. Emphasis must be placed on coaching workers, the examination of problems that occur and problem-solving using a team approach, and helping workers to achieve the skills necessary to operate the AMS. Coordination, planning and scheduling will also become more important with AMS. Finally, managers who evaluate worker and system performance must acknowledge the different approach to production characteristic to AMS. Emphasis must be placed on evaluating workers in terms of their contribution to the smooth operation of the AMS (avoiding disturbances) and their responses to problems (disturbance handling). It is also important to note how well workers coordinate their actions with other workers within the manufacturing process. Traditional piece rates, for example, will not function well within the context of AMS because they attempt to isolate a single worker's contribution to productivity. In an AMS, with its production units tightly woven into an interdependent fabric, this reductionist approach appears to be impractical.

Second-order effects occur in five areas: skill needs, personnel selection, training, personnel policies, and organizational structure. For workers, the change from conventional manufacturing to AMS requires an entirely different contribution: broad skills rather than discrete tasks, decision-making rather than manual labor, continuous growth of process-related expertise rather than duty-specific training, and proactivity in coordinating with others rather than passive performance of a particular and specific task (Welter, 1988; Sobol and Lei, 1994). This requires a management commitment to identifying skill needs and making sure that workers meet these needs, through selection and training. It also requires that management put in place personnel policies that support skill development and worker retention (experience will become very important), and organizational structures that promote coordination among workers and that foster appropriate supervision.

One important aspect of organizational change that is not strictly a structural problem has to do with fostering innovativeness. If AMS require that problem-solving skills and decision-making be more widely distributed across an organization, the organization must encourage innovative thinking about everyday problems. This is properly an aspect of organizational culture, the norms and values that workers within the organization share. The innovativeness of individual workers is related to the degree to which they perceive that the organizational culture is receptive of new ideas and is concerned about them and their careers (Hurley, 1995). Organizational culture has received widespread attention lately in the United States because of apparent failures to provide adequate attention to these areas (Adams, 1996)

9.2.3 Implementation: managing organizational change

One human impact on AMS has not yet been considered: how does one minimize disruption while implementing changes in technology and the personnel subsystem? Any change in culture may produce resistance to that change; careful AMS implementers will plan to minimize resistance and manage cultural change to the advantage of the organization as it occurs. Sun and Riis (1994) identified four stages in the implementation of a rapid response manufacturing system. These are: initiation and justification, preparation and design, installation and training, and routinization and learning. Endsley (1994) proposes that an organization must plan a five-stage approach to ensure positive adjustment to organizational change. These stages are:

1. initial decision, in which an organization decides that instituting an AMS is necessary
2. introduction, in which the deciding group educates others in the organization about the need for change and resulting benefits
3. initialization, in which workers and managers are exposed to the new methods
4. early adjustment, in which workers and managers begin operating the new system and receive their first experience with it
5. feedback, in which workers and managers help to adjust the AMS based on their experiences.

Throughout these stages, it is important that implementers appropriately involve workers and managers in decision-making and accept their feedback on the direction that the organization is taking. Providing users with a sense of ownership and a perception that their opinions count can deflect much resistance to change. In addition, this moves the organization in the cultural direction it should take to optimize the AMS.

Karwowski *et al.* (1994) proposed a strategic approach to the implementation of advanced manufacturing techniques that integrate humans and machines into a rapidly responding organization. Because this framework concentrates on company goals, development of design principles to meet the goals, implementation of organizational structures and design techniques, they labeled it GOPRIST (goals, principles, structures and techniques).

There seem to be two key concepts in these implementation approaches: first, the implementers of AMS must be aware of the· impact of technology on the way that work is done. This impact reverberates through the organization, affecting the organizational structure, the roles and tasks of managers and workers, and the skills necessary to successfully perform the work. Second, the implementers must be aware of the impact of the AMS and plan to manage the implementation process. There is a strong need to educate the members of the organization about the AMS and for accepting feedback from them. The first cultural change in an organization implementing an AMS must be within the group that is planning the change. These key people will help to set the tone for the organization during introduction, implementation and early adjustment.

9.3 MICROERGONOMICS

Normal (1995) has noted, 'As automation increases, the need to apply such [ergonomic design] principles becomes more urgent ... Workers who operated tools could view many of the parts and could see the effects of their actions. People had some hope of understanding how large machinery and small gadgets worked, because the parts were visible. The operation of modern machines and the concepts behind their design are invisible and abstract. There may be nothing to see, nothing to guide understanding. Consequently, workers know less and less about the inner workings of the systems under their control, and they are at an immediate disadvantage when trouble erupts.' 'Such alienation has startling effects: most industrial and aviation accidents today are attributed to human error.'

As Karwowski *et al.* (1994) point out, new manufacturing technologies will require more from members of the organization. Skills and abilities must be enhanced to accommodate the new approach to manufacturing. All this requires that AMS be designed with the contribution and capabilities of workers planned into the system. Designing with people in mind has been termed human-centered job design. Corbett (1990) lays out some principles for human-centered job design. Human-centered design:

1. accepts human skills and allows them to develop, rather than minimizing human contribution
2. allows users to modify their own goals and behavior
3. unites the planning, executing, and monitoring components, and avoids the extreme specialization of conventional manufacturing
4. encourages coordination and communication between workers
5. provides healthy, safe and efficient working environments.

These principles might be termed skills utilization, job flexibility, job integration, coordination, and ergonomics, respectively. Grote *et al.* (1995), in an analysis of function allocation in AMS, list five criteria:

1. coupling, which has to do with the integration of human and machine
2. process transparency/proximity, which has to do with how well the worker understands machine operations
3. decision authority
4. flexibility, which is similar to job flexibility
5. technical linkage, which has to do with how closely the machine is linked to other systems in the manufacturing process.

Unfortunately, it may be very difficult for manufacturing firms to implement human-centered job design (Fan and Gassmann, 1995). First, the sociotechnical aspects of the new technology are not so obvious as the technical, and it is easier for management to concentrate on the technical problems. These include time constraints, technical feasibility and financial and economic considerations (Lund *et al.* 1993). In this milieu, it is easy to ignore the more difficult and less easily quantifiable sociotechnical challenges. As an example, one set of guidelines for planning, purchasing and implementing flexible manufacturing systems makes no mention of how such technology might affect workers or the organization, except to say that some needed skills may not be available on the existing shop floor (Charles Stark Draper Laboratory, 1984). Second, social inertia makes it difficult to adjust to a new approach that contradicts past management practices, and human-centered approaches violate the existing culture within many organizations. Finally, workers may resent the added demands of human-centered approaches. These factors are certainly part of the reason that Pardo *et al.* (1995), in a survey of Swiss companies, found that few used a holistic, human-centered approach to manufacturing and most did not make efficient use of the skills and qualifications of their workers. This is unfortunate; certain aspects of the human-centered approach, specifically skills utilization and job flexibility, reduce downtime in AMS by as much as 70% (Jackson and Wall, 1991).

An example given by Ulich *et al.* (1990) demonstrates the problems inherent in AMS implementations driven by the technology alone without sufficient attention to sociotechnical issues. They studied an AMS that was organized around one particularly important component of the technology. The technology-centered design resulted in a system that distributed the information needed for process monitoring to locations so widely separated that it was impossible to supervise the process from a single place. Process monitoring and trouble-shooting required gathering information from several different points. As a result, the AMS was 'very susceptible to malfunction', and it prevented operators from using their skills and abilities to the fullest. Designs driven by the technology and ignoring sociotechnical issues are prone to this sort of problem. Without mutually supporting technical and sociotechnical design it is impossible to produce an optimized AMS.

Jordan (1963) observed that the allocation of functions between humans and machines hinges on the underlying principle that '[people] are flexible but cannot be depended upon to perform in a consistent manner whereas machines can be depended upon to perform consistently but they have no flexibility whatsoever.' He further pointed out that although allocation of functions depends on the ability to compare human and machine performance, the two are not comparable but are rather complementary. The three decades of development in robotics and artificial intelligence have not altered this important point. The complementary abilities of human workers and advanced manufacturing equipment should all be considered when designing methods for completing tasks in the context of advanced manufacturing systems. For AMS, developing complementarity between human workers and their machines is important because of the greater complexity inherent in AMS and desires for manufacturing flexibility. Guidelines for allocation of functions to human or machines have been developed for many years; a recent, highly detailed, approach specifically for manufacturing is a good source for methods and principles (Mital *et al.*, 1994a, b).

The traditional workstation design issues will not be much affected by AMS. Much work has already been done in studying human–machine interfaces for complex systems, and specifically, for operators of process control systems and robotic devices (Edwards, 1974; Bullinger, Korndorfer and Salvendy, 1987; Salvendy, 1987; Sanderson, 1989; Rahimi and Karwowski, 1990; and Draper, 1995). What is new will be the application of some advanced control station concepts to the shop floor. A proper ergonomic approach will be a necessary part of the AMS implementation, including appropriate function allocation, good job design, a well-done task analysis, and workstation design by persons with appropriate training and skills.

9.4 SUMMARY

This chapter introduced the reader to important sociotechnical issues in the implementation of AMS. From this, the reader should take away two key ideas: first, AMS mean more than putting different tools into the hands of workers on the shop floor. It means a fundamental restructuring of the way that work is done and of the relationships between machine operators, and between machine operators and managers. Without understanding this and planning for sociotechnical change alongside of planning for technological change, the AMS implementation will not achieve all that is hoped for it. Second, and more subtle, AMS implementation is not an event, it is a process. This is most true of the sociotechnical system. Implementers can avoid many problems with the sociotechnical system through thorough planning and by consultation with the affected members of the organization. However, it will not be possible to anticipate every problem. There will be a certain amount of evolution within the organization as workers and managers become more familiar with the AMS. Managers can deal with this by adopting an organizational culture that promotes and rewards innovation and by demonstrating that they are willing to accept feedback and act on it appropriately.

The social system that an AMS is embedded in must be as carefully planned and fully integrated as the manufacturing technology itself to achieve optimal production performance. From the top down, the organization must be designed to permit rapid responses to environmental changes. It is not sufficient to develop flexible manufacturing technology alone: the organization that uses the technology must be equally flexible. At the level of the job, the duties assigned to workers must support the rapid response capability of the technology, and be designed with flexibility in duties so that a worker can respond to the needs of the machinery without requiring appeals to specialists, as much as possible. This implies that job incumbents are sufficiently skilled, trained and motivated to respond to the needs of the technology appropriately. Finally the human–machine interfaces must provide workers with sufficient information to understand the production process and to make good decisions about how they interact with the machinery.

Finally, it is important to note that organizational design requires a champion. Someone or some group must be responsible for planning the sociotechnical changes, monitoring their implementation, and making evolutionary adjustments as necessary. This is most appropriately a role for a person or group that is part of the organization and not a consultant, because it requires a commitment to the mission success of the AMS and the manufacturing firm.

ACKNOWLEDGMENT

This work was performed at the Oak Ridge National Laboratory, managed by Lockheed Martin Energy Research Corp. for the US Department of Energy under contract DE-AC05–84OR21400. The opinions expressed are those of the authors and not necessarily those of the Oak Ridge National Laboratory, Lockheed Martin Energy Research Corp., or the US Department of Energy.

REFERENCES

Adams, S. (1996) *The Dilbert Principle: A Cubicle's-Eye View of Bosses, Meetings, Management Fads and Other Workplace Afflictions*, New York, HarperBusiness.

Berger, A. (1994) Balancing technological, organizational, and human aspects in manufacturing development. *International Journal of Human Factors in Manufacturing*, **4**,(3), 261–280.

Berniker, E. (1990) Issues and challenges to sociotechnical systems design of advanced manufacturing systems. In W. Karwowski and M. Rahimi (eds), *Ergonomics of Hybrid Automation Systems II*, Elsevier, Amsterdam, 3–9.

Bullinger, H., Korndorfer, V., and Salvendy, G. (1987) Human aspects of robotic systems. In G. Salvendy (ed.), *Handbook of Human Factors*, Wiley-Interscience, New York, 1657–1693.

Charles Stark Draper Laboratory (1984) *Flexible Manufacturing Systems Handbook*, Noyes Publications, Park Ridge, NJ.

Child, J. (1977) *Organization*, Harper and Row, New York.

Corbett, J. M. (1990) Human centred advanced manufacturing systems. From rhetoric to reality. *International Journal of Industrial Ergonomics*, **5**(1), 83–90.

Draper, J. V. (1995) Teleoperators for advanced manufacturing: applications and human factors challenges. *International Journal of Human Factors in Manufacturing*, **5**(1), 53–85.

Duffy, V., Danek, A., and Salvendy, G. (1995) A predictive model for the successful integration of concurrent engineering with people and organizational factors: based on data of 25 companies. *International Journal of Human Factors in Manufacturing*, **5**(4), 429–445.

Edwards, E. (ed.) (1974) *The Human Operator in Process Control*, Taylor and Francis, London.

Endsley, M. R. (1994) Implementation model for reducing resistance to technological change. *International Journal of Human Factors in Manufacturing*, **4**(1), 65–80.

Engstrom, T., Johansson, J. A., Jonsson, D., and Medbo, L. (1995) Empirical evaluation of the reformed assembly work at the Volvo Uddevalla plant: psychosocial effects and performance aspects. *International Journal of Industrial Ergonomics*, **16**, 293–308.

Fan, I. S., and Gassmann, R. (1995) Study of the practicalities of human centred implementation in a British manufacturing company. *Computer Integrated Manufacturing*, **8**(2), 151–154.

Gerwin, D., and Kolodny, H. (1992) *Management of Advanced Manufacturing Technology: Strategy, Organization, and Innovation*, Wiley, New York.

Gerwin, D., and Leung, T. K. (1986) The organisational impacts of flexible manufacturing systems. In T. Lupton (ed.), *Human Factors: Man, Machine and New Technology* Springer-Verlag, Berlin, 157–170.

Grote, G., Weik, S., Wafler, T., and Zolch, M. (1995) Criteria for the complementary allocation of functions in automated work systems and their use in simultaneous engineering projects. *International Journal of Industrial Ergonomics*, **16**, 367–382.

Harvey, N., and Behr, M. V. (1994) Group work in the American and German nonautomotive metal manufacturing industry. *International Journal of Human Factors in Manufacturing*, **4**(4), 345–360.

Hendrick, H. W. (1987) Organizational design. In G. Salvendy (ed.), *Handbook of Human Factors*, Wiley-Interscience, New York, 470–494.

Hendrick, H. W. (1991) Human factors in organizational design and management. *Ergonomics*, **34**, 743–756.

Hendrick, H. W. (1993) A macroergonomic perspective of ergonomics. In W. S. Marras *et al.* (eds.) *The Ergonomics of Manual Work* Taylor and Francis, London, 467–470.

Hirsch-Kreinsen, H., and Schultz-Wild, R. (1990). Implementation processes of new technologies – Management objectives and interests. *Automatica*, **26**(2), 429–433.

Horte, S. A., and Lindberg, P. (1994) Performance effects of human and organizational development and technological development. *International Journal of Human Factors in Manufacturing*, **4**(3), 243–259.

Hurley, R. F. (1995) Group culture and its effect on innovative productivity. *Journal of Engineering and Technology Management*, **12**, 57–75.

Jackson, P. R., and Wall, T. D. (1991) How does operator control enhance performance of advanced manufacturing technology? *Ergonomics*, **34**(10), 1301–1311.

Jordan, N. (1963) Allocation of functions between man and machines in automated systems. *Journal of Applied Psychology*, **47**(3), 161–165.

Karwowski, W., Salvendy, G., Badham, R., Brodner, P., Clegg, C., Hwang, S. L., Iwasawa, J., Kidd, P. T., Kobayashi, N., Koubek, R., LaMarsh, J., Nagamachi, M., Naniwada, M., Salzman, H., Seppala, P., *et al.* (1994) Integrating people, organization, and technology in advanced manufacturing: a position paper based on the joint view of industrial managers, engineers, consultants, and researchers. *International Journal of Human Factors in Manufacturing*, **4**(1), 1–19.

Kidd, P. T., and Corbett, J. M. (1988) Towards the joint social and technical design of advanced manufacturing systems. *International Journal of Industrial Ergonomics*, **2**(4), 305–313.

Leavitt, H. J. (1972) *Managerial Psychology*, University of Chicago Press, Chicago.

Lund, R. T., Bishop, A. B., Newman, A. E., and Salzman, H. (1993) *Designed to Work: Production Systems and People*, Prentice Hall, Englewood Cliffs.

Mathews, J. (1995) Organizational foundations of intelligent manufacturing systems the holonic viewpoint. *Computer Integrated Manufacturing Systems*, **8**,(4), 237–243.

Meshkati, N. (1990) Integration of workstation, job, and team structure design in complex human-machine system. In W. Karwowski and M. Rahimi (eds), *Ergonomics of Hybrid Automation Systems II*, Elsevier, Amsterdam, 59–68.

Mital, A., Motorwala, A., Kulkarni, M., Sinclair, M., and Siemieniuch, C.(1994a) Allocation of functions to human and machines in a manufacturing environment: Part I – Guidelines for the practitioner. *International Journal of Industrial Ergonomics*, **14**, 3–31.

Mital, A., Motorwala, A., Kulkarni, M., Sinclair, M., and Siemieniuch, C. (1994b) Allocation of functions to human and machines in a manufacturing environment: Part II – The scientific basis (knowledge base) for the guide. *International Journal of Industrial Ergonomics*, **14**, 33–49.

Norman, D. A. (1995) Designing the future. *Scientific American*, **273**(3), 194–198.

Pardo, O., Strohm, O., Kirsch, C., Kuark, J. K., Leder, L., Louis, E., Schilling, A., and Ulich, E. (1995) Computer-aided manufacturing systems: work-psychological aspects and empirical findings from case studies. *International Journal of Industrial Ergonomics*, **16**, 327–338.

Pilitsis, J. V., and Coringrato, E. J. (1992) Organization design and span of support. In G. Salvendy (ed.), *Handbook of Industrial Engineering* Wiley-Interscience, New York, 721–747.

Rahimi, M., and Karwowski, W. (1990) Research paradigm in human-robot interaction. *International Journal of Industrial Ergonomics*, **5**(1), 59–71.

Robertson, M. M., and Majchzark, A. (1987) Advanced manufacturing technology development: A macroergonomics perspective. *31st Annual Meeting of the Human Factors Society,* New York.

Salvendy, G. (ed.) (1987) *Handbook of Human Factors,* John Wiley, New York.

Salvendy, G., and Karwowski, W. (eds) (1994) *Design of Work and Development of Personnel in Advanced Manufacturing,* Wiley, New York.

Sanderson, P. (1989) The human planning and scheduling role in advanced manufacturing systems: an emerging human factors domain. *Human Factors,* **31**(6), 635–666.

Shipley, P. (1990) The analysis of organisations as a conceptual tool for ergonomics practitioners. In J. R. Wilson (ed.), *E. Nigel Corbett* Taylor and Francis, London, 779–797.

Sobol, M. G., and Lei, D. (1994) Environment, manufacturing technology, and embedded knowledge. *International Journal of Human Factors in Manufacturing,* **4**(2), 167–189.

Sun, H., and Riis, J. O. (1994) Organizational, technical, strategic, and managerial issues along the implementation process of advanced manufacturing technology – a general framework of implementation guide. *International Journal of Human Factors in Manufacturing,* **4**(1), 23–36.

Ulich, E., Schupbach, H., Schilling, A., and Kuark, J. K. (1990) Concepts and procedures of work psychology for the analysis, evaluation and design of advanced manufacturing systems. a case study. *International Journal of Industrial Ergonomics,* **5**(1), 47–57.

Welter, T. R. (1988) Getting set for implementation. *Industry Week,* **2**, 42–55.

CHAPTER TEN

CAD/CAM integration: an efficient approach in a fixture design framework

Daniel Noyes and Emmanuel Caillaud

10.1 INTRODUCTION

Trends in industrial design nowadays depend on the use of simultaneous design concepts with the general aims of reducing the time to market, improving quality and finally reducing costs.

Fixture methods are strongly affected by this evolution because they play a major role in the mechanism of defining of the manufacture process of the product. In particular, the fixture design, directly linked to the definition of the part, the means of manufacture, the definition of the tools used and their commitment, is in the heart of integrated design developments. Our presentation is placed in the field of the CAD/CAM integration. Our contribution to the CAD/CAM integration takes place at the level of the fixture definition and its integration in the cycle of the part design.

The objectives of our work are: at the fixture level to structure the design process and the realization of the fixture using integrated design concepts; and at the part level to propose this fixture design framework as the support for the cooperation between the activities of the part design.

The presentation is organized in three parts. First, we review the context of simultaneous engineering in which our developments are included. Thus, we expose:

- the methods associated with the simultaneous engineering concepts
- the fixture role in the realization of these concepts at the level of the part manufacture
- the importance of the mastering of corresponding technological knowledge.

Then we propose a generic process adapted to handle technological knowledge during the elaboration and realization of any product. The process is called KHAP; it is based on the definition of mechanisms of knowledge and reasoning acquisition, modeling and validation and, finally, their implementation.

The developed key points are:

- the knowledge acquisition, a capture sequence that relies on a first model is proposed; this model is propitious to the communication and to a broad capture and a first form of ratification; the characteristics of an approach by 'immersion' in enterprise are exposed
- the explicit knowledge representation, supports adapted to analyze levels of accuracy and the completeness of this knowledge are introduced; as technological knowledge has a vague and incomplete character, the modes of representation must allow a high degree of flexibility in modifying or complementing
- the validation of the knowledge, sequences involving both tests and confrontations with the expert's knowledge are proposed; the notion of confidence level in the validation is considered.

Finally, we present the application of this methodology to model the knowledge in the method department for the fixture design in manufacturing production.

Our developments concern:

- the definition of the design process of the fixture using a formalism that underlines the links between the part design, the process planning definition and the fixture design
- the procedure for fixture design based on fixture designers' rules and the expression of these rules
- its implementation by means of an expert system which provides a helpful tool to the fixture design.

10.2 INTEGRATED DESIGN AND ITS ATTRIBUTES

The general objectives of the integrated design are to reduce the time to market of the product, to improve the quality of the product and to reduce costs.

The underlying concepts of the integrated design are the design for manufacturing (DFM), the concurrent engineering and the design for reactivity (DFR), the last of which has been classically less developed. The corresponding concepts associate simple principles in their definition which are nonetheless delicate to implement. We summarize these ideas below.

10.2.1 Integrated design approaches

Design for manufacturing

The DFM's objective is to reduce the complexity of products to facilitate their realization (and assembly) to reduce the cycle of design-manufacture and associated costs. The principles used allow robust decisions to be made at each step of the design, by the knowledge of the functioning conditions of downstream activities, at each decision level. A collaboration exists between the steps; it consists of giving to each decision-making step a more or less aggregated knowledge of the functioning of the downstream level(s) (constraints, trade rules, . . .) to allow it to establish, knowing all the facts, a compatible solution with the expectations of downstream steps.

Two forms of action are possible in this context:

- anticipation of the constraints of the downstream step by taking them into account at the level of elaboration of the upstream step's solution
- construction of a minimal solution at the level of the upstream step by committing only truly indispensable decisions at its activation time to relax constraints imposed on the downstream steps (Brun, 1994).

Both approaches aim to eliminate the downstream step's calling into question the already realized steps, but each approach has to avoid particular pitfalls. In the first case, the evident risk is an important overspecification of the downstream activity that hampers the development of its solution; in the second, the main problem is an insufficient formalization of the solution at each level by transfer of decisions and thus concentrating a lot of key decisions at the level of the last steps.

Concurrent engineering

The simultaneous design has the same objectives of costs and time to market reduction and improvement in the quality of the product by the placement in parallel of activities which have been traditionally placed in series (Sohlenius, 1992). It is based on three principles: parallelism, competition and integration.

The placement of activities in parallel imposes the decomposition of the design – manufacture activity into projects. The competition among several solutions for each project allows one to choose the best of them. The integration of activities allows one to take into account the constraints of each 'actor' and to obtain an optimal solution.

The integrated design principles are entirely compatible with the DFM; their application changes the classical sequence of activities into a network of simultaneous activities.

Design for reactivity

The basic principle of the DFR is to increase, for each step of a decision, the set of degrees of freedom forming its decision framework (Aldanondo *et al.*, 1995). This situation is established by developing, for each upstream step, a set of possible solutions in substitution for a unique one; two effects are induced: a relaxation of the transmitted constraints to the downstream step; and a widening of the decision framework of the downstream step. The result is an improvement in the reactivity of the downstream activities that are faced with the situations to manage. This is due to a larger available set from which elements of solutions can be extracted.

In these conditions, any perturbation intervening at any level must be able to be brought down locally without canceling the upstream decision with a failure report. However, it is evident that this approach relies on well-known frameworks for the downstream decisions and mechanisms implemented at these levels to offer them true degrees of freedom. On this point a difficulty exists: the capture of expertise allows direct access only to the knowledge committed in traditional decision frameworks (rules of trade, mechanisms of reasoning) and an abstraction effort is solicited from the expert to define his behavior in a different framework.

The DFR is entirely compatible with the other concepts of integrated design of which it strengthens the efficiency.

10.2.2 Integrated design scheme and fixture cycle

In manufacturing production, the realization of the first part is the ultimate form of validation of the product and the process by which it is obtained. In the end, it alone allows one to verify the quality of the reply to the client's expectations (explicit and implicit), to validate the process of production, the means necessary for its implementation and, in particular, the fixture.

The fixture is a real physical interface between the part and the production machine and is in the heart of the process of production. Often an important delay is generated by its design and its realization. The fixture can thus be an obstacle to the reactivity in the general process of the product design.

We summarize in Fig. 10.1 the possible parallel functions in a simultaneous design approach which serve to decrease the time needed to obtain the first part (the links between functions illustrate usual relationships between these functions).

Implementing simultaneous design makes it essential to master the knowledge used by the actor at each step to extract from it the necessary knowledge for the other actors. In any case, an application of integrated

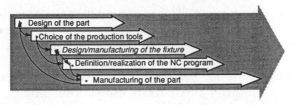

Fig. 10.1 Dependencies and parallelism of the part design functions.

design requires one to make part of the knowledge common. At this point, a basic problem is to extract and to model the technical and technological knowledge underlying the design and method departments' activities. We will study this problem in the following sections.

10.3 KNOWLEDGE MODELING IN DESIGN

The presented methodology aims to model knowledge in design in industrial manufacture production. We propose a progressive process for the technological knowledge acquisition, modeling and validation. Our approach is based on: the same principles used in the 'classical' methodologies of knowledge engineering; and additional 'specific' principles to handle technological knowledge which form a set called KHAP (technical knowledge handling additional principles).

We propose with KHAP a modeling type dedicated to technological knowledge, thus privileging the quality of the relationship with the experts. We propose also an acquisition by immersion in enterprise, a modeling of the area of study, a procedural structuring of the followed process, a structured expression of rules and data and a validation by an exploitation of knowledge using an expert system.

This methodology consists of three main steps: the acquisition, modeling and validation of knowledge. It is based on the increasingly refined knowledge analysis and on cycles of modeling validation with the concerned actors.

10.3.1 Knowledge acquisition

Technological knowledge extraction is delicate because it relies on an inexplicit knowledge which is therefore purely expert (based on experience and not on defined procedures). The knowledge engineer must explicate and model the knowledge. Then, this structured knowledge can be used in an application. In fact, if experts use their know-how to solve problems, they cannot necessarily easily explain their way of thinking. To collect knowledge efficiently, a modeling phase for the process followed is necessary. This modeling may have two objectives: to capture the knowledge to obtain the same result as the expert or to imitate scrupulously the expert's way of reasoning. Knowledge extraction implies a direct relationship between the knowledge engineer who must capture the know-how and the expert. The quality of the result depends on the quality of this human relationship. This situation implies a modeling of the knowledge, which requires a great deal of effort. Among the main difficulties, we underline first the modeling and then the validation of the unexpressed knowledge.

Let us consider two complete and recent methodologies for knowledge extraction, which are MACAO and KADS among the several existing ones to set up the basic principles.

MACAO is a general method to support knowledge extraction (Aussenac, 1989). The aim of MACAO is to build a conceptual complete model of the know-how from the capture of the knowledge in various ways (interviews, analysis of data, . . .). MACAO is applied in three main steps:

- extraction and modeling of knowledge from data in the domain
- modeling of the domain and the process followed by the expert
- knowledge extraction of the missing rules based on the model of the reasoning method.

KADS proposes a top-down process of knowledge extraction based on the transformation of pre-defined models (Tansley *et al.*, 1993). The aim of KADS is to propose a methodology for knowledge extraction starting with the modeling of the domain all the way to the implementation of the full system. KADS is made of the following steps:

- a study of the limits of the domain
- an extraction and modeling of the data
- a modeling of the domain
- a top-down modeling of the reasoning method based generic models
- an extraction of the knowledge constituting the steps of the reasoning and prototyping model.

These two methodologies make the link between the psychological approach to human knowledge and the extraction of the necessary knowledge to develop a knowledge-based system.

The approach that we present resumes these principles and proposes modeling type, adapted to technological knowledge, which improve the quality of the exchanges with the experts.

The primary KHAP's principles which correspond to the knowledge acquisition are shown in Fig. 10.2.

Step S1: to acquire and model the knowledge, the domain must first be analyzed. The domain modeling must precede the extraction, modeling and validation of the experts' rules. This work allows one:

- to define the limits of the field of knowledge to be acquired, to identify the experts and to capture part of this knowledge
- to analyze the firm's organization with regard to the studied know-how and to identify ways of integration of the know-how with other activities of the firm
- to structure the field by a functional analysis
- to prepare the implementation of concurrent engineering principles in identifying links and interrelations between actions.

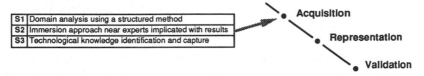

Fig. 10.2 KHAP principles for knowledge acquisition.

Respecting a formalism provides an aid for the domain validation. Several formalisms can be used for this. We have used the SADT™ formalism (Softech, 1976). An SADT™ model is a graphical representation of the hierarchical structure of a system which clearly reveals the relationship between the different elements of the system. In this way, it is possible to understand a complex system in a complete and precise manner and to communicate.

An SADT™ model can be validated. Our choice for this formalism is mainly due to the possibility of validation by the firms' actors. The modeling is realized with actigrams to give importance to the functions implied and their relationships. This allows one to precisely situate the studied process in its environment and to structure the vocabulary of the knowledge.

Step S2: knowledge extraction must be made by experts in the domain who are directly interested in the expected results.

We recommend an investigation by 'immersion' in the firm to directly meet the owners of the know-how to the model and also to the actors in relationship with this know-how. Meeting persons of high hierarchical levels provides data on the structure and organization of the departments. People who are closer to the realization can provide data on the relationships with other departments and on the functions fulfilled by the department.

This work allows several points of view and reveals the necessary links for integration (contributing to the application of simultaneous engineering).

Finally, we recommend cross-checking the analysis of the expert's reasoning with that of his behavior in real situations.

Step S3: knowledge capture starts with these first steps. It is made in different ways (interviews, analysis of documents,...) and we can guarantee its completeness (of the covered fields but not of the knowledge itself) by comparison with the domain analysis. This knowledge capture corresponds to the first approach of the experts' reasoning method. We distinguish the technological knowledge from that of the domain. We also consider that any field of technological knowledge classically combines generic knowledge, including the basic theory of the domain, with the real know-how. We are specially interested in this last knowledge because it is the real know-how that is used for the specific reasoning.

10.3.2 Knowledge representation

The reasoning used by the expert must be defined to structure the underlying rules. The corresponding KHAP's principles for knowledge representation are as shown in Fig. 10.3.

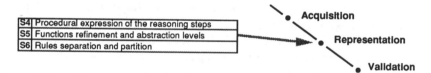

Fig. 10.3   KHAP principles for knowledge representation.

Step S4: we propose a procedural expression for the steps of the reasoning. Note that the expert generally follows a reasoning by analogy. He compares the initial conditions of the problem to the initial conditions already met. He then deduces the solution to the problem with regard to a solution already realized. This predefined basic solution is then modified gradually, taking into account the specific constraints of the studied problem (due to the differences between the two examples). The expert generates only a single solution which is progressively modified during the design.

This basic solution is often deduced from a small set of cases which are used as references by the expert. These references correspond to the most recent cases and the most representative cases. The expert will try to modify the basic solution as little as possible in spite of the differences between this solution and the studied problem. Moreover, the expert designs directly out of habit, according to the available physical elements and not according to the requirements of the product in question. Thus, the analysis of the design process is difficult.

Note that reasoning by generation (followed by expert systems) consists of generating a new solution from the initial data without comparing it to other situations already encountered. The know-how is not translated by basic solutions linked with basic examples but by rules of design structured in a procedure.

We think that the modeling of the knowledge by a procedure which generates of solutions (and not by analogy) allows one to express explicitly the knowledge for design. The approach by analogy hides the knowledge used because the solution is not generated but deduced.

Step S5: the problem studied (analysis of the design process) brings another dimension to knowledge: the refinement of the functions defined. We translate this refinement with several levels of abstraction for each function. The design process leads to refinement of each function definition and to the allowance of a progression from one function to the another.

We define three levels of abstraction: functional, technological and physical. These levels can be used for the structuring of any process of product. One difficulty in representing knowledge by design is that the expert (designer) expresses only the final solution (physical solution) and does not express the progressive refinement of the functions. The expert neglects the identification of the functions and primary technological choices and instead focuses on the size of the physical elements. To avoid this difficulty, we propose a decomposition of the design reasoning into two steps: the definition of the virtual product (functional and techno-logical solution) and the definition of the physical product. At the functional level, functions and their interrelations are defined with respect to the problem requirements and the applicable rules. At the technological level, the technological solutions which fulfill these functions are detailed. At the physical level, the physical elements that answer to these specifications are defined. This results in a structured representation of the expert's reasoning including the process used as a framework for the use of the expert's rules.

Step S6: for each step of the process, the objective, the necessary data and the outputs must be defined. The constitutive rules of each step are then listed as production rules.

The nature of the objectives does not correspond to a binary answer (objective reached or not) but to the expression of levels of satisfaction in their fulfillment. This makes necessary rules evaluation for these levels of satisfaction but also selection rules according to the values reached by these levels.

The needed set of data often consist of: basic data directly available at the level of the problem definition, elaborate data resulting from 'translation' (aggregation, composition, refinement,. . .) of the basic data, implying simple rules or an expert know-how.

The outputs result from the commitment to the expert's rules. The expressed knowledge often looks like the logical scheme

$$[\text{input data} \xrightarrow{\text{rule (or set of rules)}} \text{result}]$$

whereas the real scheme implies other methods, unconciously studied and given by the expert. Our objective is to collect this knowledge.

Therefore, we define a typology of the rules and data to a precise analysis of the engaged knowledge. We have defined two main types of data: rough data (type DI) and elaborate data from rough data (type DII). We can distinguish four types of rule bases:

- rule bases to define necessary data for the resolution of the problem from the available data (type RI)
- rule bases to define solutions from the data (type RII)
- rule bases to evaluate partial or global solutions (type RIII)
- rule bases to select acceptable solutions (type RIV) from the results of the evaluation.

Generally, experts mix definition, evaluation and selection according to our previous comment. Only type RII rules are easily accessible directly. The different types of rule are presented in Fig. 10.4. This step of the methodology allows one to structure the process followed by the expert and to define the rules on which the reasoning is based.

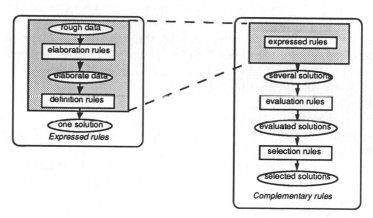

Fig. 10.4   Different types of rule.

10.3.3 Knowledge validation

The acquired knowledge must be validated according to two complementary axes: verification of the completeness and the coherence of the rules and validation of the semantic by the experts.

The validation must allow ratification of the acquired knowledge in the domain at the level of the design procedure but also at the level of the results provided by the use of the rules. The design procedure must be validated by interviewing the experts. The experts must validate the different steps and also the logical links between them.

As the capture of knowledge is difficult (rules non-formalized, contradictory and often non-expressed), we look for a way to simulate this knowledge to modify and increase it. The knowledge must be rigorously structured to allow a complement and a local validation without resulting in repercussions to the whole procedure.

According to the above remarks, the KHAP principles that we have defined for knowledge validation are proposed in Fig. 10.5.

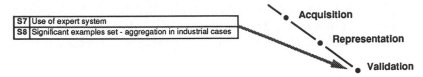

Fig. 10.5 KHAP principles for knowledge validation.

Step S7: we propose a validation of the knowledge through its exploitation by an expert system (see, for example, Barr *et al.* (1981) and Farreny *et al.* (1987) for the principles and the features of expert systems). This type of tool leads to a strict formalization of the knowledge and offers a form of expression and use understandable by the experts. The separation of the inference motor and the knowledge base allows the validation of the motor before its effects on the knowledge base.

We propose here a structuring of the rules in independent rule bases managed by a control base using the structure of the procedure formalized. Let us note that the expert system may become an assistance tool.

Step S8: the validation is made by use of the expert system in several examples. The provided results will be approved if they are considered acceptable by the experts. This validation must be realized by using examples that are representative of a part of the domain. The validated domain is made of the part of the domain which concerns the areas covered by these examples. Each example is used for targeted interviews with the experts which validate the results provided by the expert system. The interviews allow us to confirm certain parts of the knowledge and to modify and to acquire other parts of the knowledge.

At first, we recommend the use of the expert system on examples which have well-defined particular difficulties. Then the next step is to validate the expert system in industrial cases. This allows the integration of the different levels of knowledge implication and their global validation. We understand therefore that if the use of a software allows us to generate several evaluated solutions, the reasoning is not the same as the expert. The main differences between the two approaches are: the generation of several solutions, and the multi-criteria analysis of the different possibilities.

The differences between the approach followed by the expert and the one used by the expert system are the basis of the difficulties of validation by experts. How can we validate the expert system design process if the experts do not follow the same process? Even if the design reasoning can be globally validated by the experts, this is not the case for the particular steps in the management of solutions and their multi-criteria evaluation. Thus, there is an obvious limit to validation.

Experts are particularly interested in the validation on complex examples. It is very difficult to make them work on usual situations. The basic rules can therefore not be included in a validation by exchange with the designer. However, the evaluation by the experts of the final solutions allows one to make a decision on the acceptability of the proposed design procedure.

10.4 KNOWLEDGE MODELING FOR FIXTURE DESIGN

We propose a structured analysis of the knowledge implicated in the fixture design and the exploitation of the corresponding results using an expert system to promote reactivity with an expert-system for fixture design.

The fixture must assume the mechanical link between the part and the machine-tool for one or more machining operations and must allow the part deliverance to have the required quality and lower cost according to (Dietrich *et al.* 1981).

The fixture design is a determinant step to the time to market. A lot of work has been concerned with this step (see the bibliographical synthesis presented by Hargrove *et al.* (1994), but some works are dedicated to partial aspects of the design (precision of the placement in position, maintaining forces, . . .). The bibliography does not offer a real proposal for a procedure for fixture design.

The proposed results concern the definition of an assistant tool for the fixture design. It must permit a diminution of the spending time and the definition of relevant information. It must also facilitate the fixture design integration into the part design and, moreover, allow a simultaneous development of several solutions. The goal of this tool is to use the expert knowledge and computer power and not to imitate the expert's behavior.

We present in the following paragraphs the results of the use of the KHAP principles for fixture design. We will develop the steps summarized in Fig. 10.6.

10.4.1 Knowledge acquisition (S1, S2 and S3)

A fundamental prerequisite is to precisely situate the activities concerning the steps of the fixture life cycle in the organization of the enterprise:

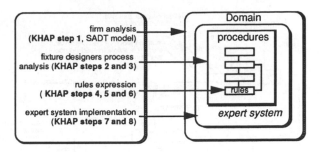

Fig. 10.6 Methodology steps.

links between design department, method department and workshop. In firms it is easier to study the product life cycle than that of the fixture. It is also necessary to situate the fixture in the firm and to study its production process according to the production process of the product. This work and the capture of expertise rules have been accomplished in different firms of aeronautic industry chosen for their representativeness in the area of mechanical manufacture of small and medium series.

We have defined a generic functional model following SADT™ formalism to delimit the area of study. This model is detailed by Caillaud *et al.* (1993). The interviewed persons consisted mainly of fixture designers (at different hierarchical levels) and other persons who intervene in the fixture environment (programmer of numerical control machine-tool, assistant and controller).

To precisely analyze the situation of the fixture, the activity of the definition of the production process has been split into two subactivities: the definition of the production process and the machining means on one side and the realization of the machining means on the other side. We present in Fig. 10.7 the diagram corresponding to 'To prepare the production'. The diagram consists of three functions: 'To define the machining set-up', 'To define the tools', 'To define the fixture'. We particularly study this last function.

The realization of each machining means is independent of the others. However, the realization of the NC program imposes knowledge of the definitions of the implied machining means.

Note that, for the general model, the supports of the function are not always defined because some of them depend on the firm's organization into departments (these supports are presented as broken arrows pointing to the bottoms of the boxes). The general model has been realized by taking into account the particularities of the models of the different firms studied. It has been easily verified that each particular model can take place in the general model. This model can be used to study the organization of the firm with respect to the fixture process and to allow one to answer questions such as which department is in charge

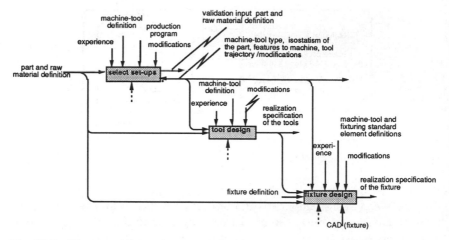

Fig. 10.7   Diagram A2, to prepare production.

of the identified functions, and how are the data exchanged between functions? The answers to these questions are directly linked to the reactivity of the firm.

The development of this model allowed us to identify experts and to acquire part of their experience. This analysis is integrated in the step of knowledge acquisition and it contributes to the precise definition of the area of our study.

10.4.2 Knowledge representation (S4, S5 and S6)

We propose a fixture design process based on the analysis of the mechanisms used by experts. It concerns the processing of prismatic parts. The corresponding trade rules used in the firms are not formalized and sometimes are contradictory, and furthermore their utilization is empirical. Thus, we have defined a process which is a framework of the expression of the expertise.

The procedure presented includes the expert's rules for fixture design that we have formalized. This procedure is original because prior papers on this subject (Ingrand 1987, Pham *et al.* 1990, Boerma *et al.* 1980) present only a partial point of view of this problem.

Let us remember that the main functions of the fixture are part positioning, clamping and supporting. The capture of the fixture designer's knowledge was delicate because they tend to focus on the size of the physical solutions. They do not pay attention to the progressive refinement of the functions linking the characteristics of the part to the requirements of the fixture. To tackle this problem, we define a virtual fixture as the set of requirements of the fixture. The physical fixture can then be defined from these functionalities. This distinction between the

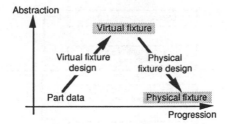

Fig. 10.8 Fixture design cycle.

virtual and the physical fixture is the result of a modification of the abstraction level for the realization of the solution (Fig. 10.8).

Twelve steps have been identified and validated along with the procedure representing their organization. This model of the fixture design process presents the classical sequence of fixture functions (positioning, clamping and supporting) but it distinguishes three levels of abstraction: functional, technological and physical. The virtual fixture design involves the first two levels and the physical fixture involves the last (Fig. 10.9).

At the functional level, we choose surfaces then zones of the part where the fixture function will be applied. At the technological level, we list the solution in terms of type, number and position of contacts. At the physical level, we choose and situate the physical element which satisfy the specifications. A modular approach for building the fixture could be considered at this third level (Hoffman 1987). Another function completes this set, the link with the machine-tool (connection between elements, fixture positioning and holding of the fixture relative to the pallet of the machine-tool). This function only exists on the physical level.

The steps of the process are not independent of each other. The realization of each function influences one's ability to realize the other functions.

Dependences and loops between steps have been revealed. Several types of loop between steps exist. If a solution is proposed at a step, it can be completely revised so long as the conditions required for the following steps are not fulfilled.

The diagram in Fig. 10.9 gives the sequence of levels for the different functions that must be achieved.

The proposed process is a simple and efficient formalization of the reasoning that one needs to follow to design a fixture. This process constitutes a framework for the utilization of trade rules in the fixture design allowing a refinement of the functions level by level.

Note here that the realization of the functions must preserve the free circulation of tools, a good lubrication and a correct chip evacuation. These rules belong to the physical level.

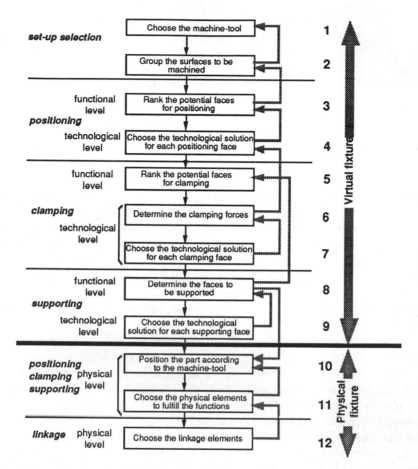

Fig. 10.9 Procedure for fixture design.

We have expressed rules only at the technological and functional levels. The definition of solutions at the physical level depends on the technical solutions' choices. It is strongly linked to the firm and the types of production it realizes.

To illustrate this formalism, let us consider the example of step 3. To describe this step, we defined four files: the first three to classify, respectively, the available faces for principal positioning, the available faces for secondary positioning and the faces available for tertiary positioning; and the fourth to evaluate the positioning system. The files have a common structure that is detailed in Fig. 10.10.

Reactivity must be facilitated by way of expression of the part data. For fixture design, the part model must not limit the design of a single solution deduced from a unique translation of the part model.

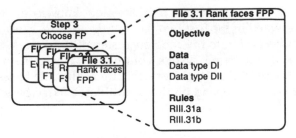

Fig. 10.10   Structure of the files of the step 3.

Several papers deal with product modeling oriented to facilitate the definition of the production process (Brun 1994). Numerous models of the part based on features have been developed (see the survey (Salomon *et al.* 1993)). In our work, we have kept this principle of decomposition into features. We have modeled the part into faces. Each face has a type and particular characteristics by which the decisions for fixture design will be made.

Let us present as an illustration how one of the essential characteristics is treated: the dimensioning. The dimensioning is converted into the accuracy of the face according to the faces to be machined in the specific set-up (Boerma 1988). Moreover, we associate some qualitative characteristics to the faces to cope with the unknown (at this level of design) cutting conditions (tool trajectories, cutting parameters) which influence the cutting forces and consequently the tool design.

Thus, all the characteristics of the part are expressed as intrinsic characteristics of the studied face. The structuring of the part into faces allows one to choose the relevant face(s) for each function of the fixture. Each function is supported by one or several faces. In this way, the characteristics of the faces allow one to define more precisely the functions of the fixture.

This approach permits the link between the characteristics of the part and the fixture function interfaces with the part. Futhermore, it allows us to develop several solutions because it releases itself from the classical dimensioning which usually guides the fixture designer to a single solution.

10.4.3 Knowledge validation (S7 and S8)

We have chosen to develop our solution on the NEXPERT™ expert system shell. We chose NEXPERT because of its properties of possible integration into CAD/CAM software, and the portability and distribution in the firms. 200 rules are implemented in a dedicated expert system called SEACMU (système d'Aide A la Conception de Montages d'Usinage). The structuring of the implemented knowledge respects the typology presented in the first two parts of the text, i.e. rules are listed by distinct basis according to the concerned step of the process and the rule type. A special rule-based system controls the engagement of the rule bases. We present in Fig. 10.11 the structure of the rule bases. Note here that the shared data depend on the control knowledge base (file 0.0).

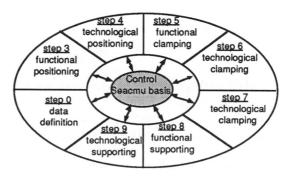

Fig. 10.11   Knowledge bases organization.

The design process and the identified rules have been validated by interviewing experts in the area and implying the expert system in several examples (the provided results will be approved if they are considered acceptable by the experts). The examples were industrial cases relative to usual problems and specific problems with particular difficulties. The primary interviews lead us to modify and to acquire complementary knowledge. The following ones allow us to progressively confirm the whole established knowledge.

Another view of the knowledge organization is seen in Fig. 10.12. This scheme is based on the previous defined typology relative to data and rules. This scheme illustrates two main points.

The set of rules RI allows us to define elaborate data (data of DII type) from the rough data (data of DI type) obtained from CAD or CAM software (we also distinguish two types of DII data which are, respectively, DIIa and DIIb types according to the way that they are obtained by a simple geometric transformation or by experts' rules). We will use these RI rules to make a link between the CAD-CAM system and SEACMU.

The sets of rules RI and RII, explicitly expressed by the experts, are not sufficient to manage concurrent solutions. The additive RIII and RIV rules allow one to evaluate these solutions and to make a selection. These two types of rule permit one to distinguish between the definition of a solution and its evaluation and selection to lay down a strict formalization of the design process. They also allow the simultaneous

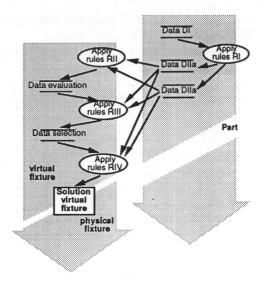

Fig. 10.12 Typology of rules and data.

development of several fixture solutions. In fact, SEACMU provides several solutions of virtual fixtures. The virtual fixture solution defines for each fixture function: the face(s) supporting the function, and the characteristics of the contacts between the part and the fixture. These contacts are defined by their type, their number and their distribution. From these requirements, the 'physical' fixture can be defined.

For the definition of the physical fixture, some analogous rules can be established. The possibility of the simultaneous development of several physical solutions contributes to the reactivity. The realization of the fixtures on the basis of modular elements also involves this reactivity.

10.5 CONCLUSION

We have presented a methodology for the acquisition, representation and validation of technological knowledge. We have exposed the results in the application of the methodology to the fixture design and the followed approach seems particularly adapted to this case.

Let us review the main steps. The design process was structured in the form of an approach sequencing the different steps. The rules forming these steps were classified into one typology. The data necessary for the realization were also outlined. An expert system using this structure of knowledge was realized with NEXPERT. This expert system allows one to develop several solutions of virtual fixtures rapidly and efficiently. Three main characteristics contribute to the reactivity:

- the expression of the part data which favors the development of several solutions of fixtures
- the modeling of the know-how for fixture design, which eases the link with the part CAD-CAM system
- the formalization of the know-how for fixture design in different types of rules adapted to the simultaneous design of several virtual fixtures.

This explicit structuring permits us to integrate the results following the concepts of integrated design. The defined methodology and its supports constitute an efficient tool to aid in the design of fixture design which is a key step in the CAD/CAM integration. This step can be integrated at the CAD/CAM process of the product, and thus it may contribute to the implementation of the integrated design concepts.

ACKNOWLEDGMENTS

This chapter presents parts of the results of a study supported by Association Nationale de la Recherche Technique, Ministère de la Recherche et de la Technologie through a contract CIFRE between Ecole Nationale d'Ingénieurs de Tarbes (ENIT) and Centre Technique des Industries Mécaniques (CETIM), contract 183–90. CETIM was involved in the BRITE-EURAM project IDEFIX 3480 on modular fixtures design. We thank particularly MM. Anglerot and Padilla (CETIM) for their support to this project.

REFERENCES

Aldanondo, M. *et al.* (1995) Une structure d'activités en conception intégrée pour l'amélioration de la réactivité de l'atelier. *CIGI' 95–La productivité dans un monde sans frontières*, Montréal.
Aussenac, N. (1989) Conception d'une méthodologie et d'un outil d'acquisition de connaissances expertes. Thèse de doctorat en intelligence artificielle, Université Paul Sabatier, Toulouse.
Barr, A. *et al.* (1981) *The Handbook of Artificial Intelligence*, William Kaufman Inc.
Boerma, J. R. *et al.* (1989) Fixture design with FIXES: the automatic selection of positioning, clamping and support features for prismatic parts. *Annals of the CIRP*, **38**, 399–402.
Brun, J. M. (1994) Un modèle produit: pour quoi, pour qui? *Modélisation géométrique et technologique en conception de produits. TEC'94*, Grenoble, PRIMECA.
Caillaud, E. *et al.* (1993) Towards a simultaneous design of the part and the fixture. *Proceedings of APMS*, Athens, IFIP, Elsevier, 243–250.
Dietrich, R. *et al.* (1981) Précis de méthodes d'usinage – méthodologie, production et normalisation. *AFNOR-NATHAN*.
Farreny, H. *et al.* (1987) *Eléments d'Intelligence Artificielle*, Hermès.

Hargrove, S. K. *et al.* (1994) Computer-aided fixture design: a review. *International Journal of Production Research*, **32**, 733–753.

Hoffman, E. G. (1987) *Modular Fixturing*, Manufacturing Technology Press.

Ingrand, F. (1987) Inférence de formes à partir de fonctions, application à la conception de montages d'usinage. Thèse de troisième cycle, INP, Grenoble.

Montalban, M. (1987) Prise en compte de spécifications en ingénierie, application aux systèmes experts de conception. Thèse de doctorat de science, spécialité informatique, Université de Nice.

Mony, C. (1994) DFM: enjeux, tendances et état de l'art. *RAPA*, **7**, 15–25.

Pham, D. T. *et al.* (1990) Autofix, an expert CAD system for jigs and fixtures. *International Journal of Machine Tools Manufacturing*, **30**, 403–411.

Salomon, O. W. *et al.* (1993) Review of researh in feature-based design, *Journal of Manufacturing Systems*, **12**(2), 113–132.

Softech Inc. (1976) SADT™, author guide.

Sohlenius, G. (1992) Concurrent engineering. *Annals of the CIRP*, **41**, 645–655.

Tansley, D. S. W. *et al.* (1993) *Knowledge-based Systems Analysis and Design, a KADS Developper's Handbook*, Prentice Hall.

CHAPTER ELEVEN

Fixtureless NC machining for rapid prototyping

Hugh Jack

11.1 INTRODUCTION

Recent developments in stereolithography and other related processes have challenged the role of cutting machine tools in producing prototype parts. The newer technologies propose faster turnaround at lower costs per part, but the initial expenditure and material costs for these processes are high and the variety of materials is limited. Many manufactures are not willing to experiment with newer methods; they would benefit from parts produced on conventional NC milling machines, but using the philosophy of the rapid prototyping processes.

A recently developed process uses composite layers of thermoset and thermoplastic polymers, among other possible combinations. These categories of materials are common and can be purchased at any local hardware store. Essentially a layer of thermoplastic material is heated, poured and allowed to harden. A cavity is then milled, and filled with the thermosetting material. This process of making layers is repeated above the bottom layer(s) until a complete thermoset shape is enclosed in a block of thermoplastic. This material is then melted away and the solid part remains. As an example, a ball could be made in two layers.

11.2 THE STATE OF INDUSTRY

The range of tools available for design engineers is steadily growing. Computer-aided engineering tools such as finite element analysis allow for estimates of stress/deformation, heat, magnetic fields, etc. Using these the designer is able to test virtual prototypes at lower cost and higher speed. Even with these tools the building of prototypes is sometimes necessary, but using the conventional machine shop can be expensive and slow. Prototype parts can be used to test look and feel, test assembly techniques, test function and build tooling. New techniques for building prototypes include numerical control machining and stereolithography. These processes focus mainly on decreasing part production times, and as a result they also become more economical.

The family of processes that are often considered to be the most successful rapid prototyping techniques are called free form fabricators. A very simplistic list of the characteristics of free form fabricators is as follows:

- complex geometries, including assemblies and hidden features
- short fabrication times, typically hours or days
- parts for testing, at worst parts will seem like a brittle plastic
- the processes are automatic.

Free form techniques tend to have little shape preference when building, and are capable of producing parts with a topological genus greater than one (e.g. assembled chain links).

Of all the rapid prototyping tools available, we can dichotomize the types of processes into additive and subtractive ones [6]. Subtractive processes form a shape by removing small quantities of metal from a larger piece. The conventional machining techniques (e.g. turning) would be classified as subtractive. On the other hand, additive processes build parts by adding material to a based part. The newer free form fabrication techniques are almost exclusively additive. Until recently the subtractive processes were the only major alternative available. Some manufacturers have begun to use the additive techniques. Some of the reasons for adopting these technologies are as follows:

- fast part production, very short turnaround times
- CAD models used directly
- reduced need for skilled supervision
- to reduce prototyping costs.

The additive technologies are quickly finding their place in the engineer's office [3], within the automotive and aerospace industries, for example [6]. Outside the typical applications, they have been used for medical prosthetics [1], surgery models [17] and molecular models [14].

11.2.1 Additive processes

Additive processes incrementally build a part by adding small amounts of material to a base part. This material is typically added in thin planar layers, or extruded along paths, with a thickness of 0.03–0.5 mm. Some of the well known techniques to employ these methods include:

SLA, stereolithography apparatus
SGC, solid ground curing
LOM, laminate object manufacturing
SLS, selective laser sintering
BPM, ballistic particle manufacturing [5]
MJM, multi-jet modeling
FDM, fused deposition modeling
Sander's prototype
DDP, droplet deposition on powder
MD*, mask deposition.

These processes involve a number of unique approaches (except where stated, all other references in this section are from [6]). SLA and SGC both use lasers and photopolymers to build parts. In each case thin layers of the photopolymer are incrementally added in a liquid state. With STL each layer is hardened by a laser that scans across the surface in a combination of raster and vector patterns. With STL, supports might be needed to make the part stable. With SGC, the polymer is partially occluded by a mask (for the new layer) and is exposed under UV light. The main differences between the techniques is that, whereas STL uses a large vat of polymer that the part is lowered into as it is produced, SGC removes unhardened polymer and replaces it with wax; as a result the supports needed in STL are eliminated. In both cases the materials are somewhat expensive, and the machine costs are relatively high. 3D Systems Inc. offers a system with a work volume of 0.5 m by 0.5 m by 0.6 m with an accuracy of 0.05 mm (accuracy is crudely chosen as the larger of resolution and repeatability) at a cost of $US450 thousand. Cubital Ltd. offers a machine with a work volume of 0.35 m by 0.5 m by 0.5 m with an accuracy of 0.15 mm at a cost of $US550 thousand.

LOM parts are built by adding material from rolls of material, and then they are cut by lasers using vector patterns. Typical materials include plastics, papers, etc. One of the major problems with this technique is that during processing excess material is left to support the part, but in post-processing this becomes difficult to remove. Also, any sloping surfaces not close to vertical cause problems. Material costs for this process can be very low, but the equipment costs can still be relatively high. Helisys Inc. sells a machine with a work volume of 0.56 m by 0.81 m by 0.5 m with an accuracy of 0.25 mm at a cost of $US180 thousand.

SLS uses fine powders that are added in layers and sintered together using a laser. This process can use a wide variety of materials, but at this point they are generally low-temperature plastics. Material costs for this process are generally low, but the equipment cost is high. The DTM corporation sells a machine that has a work volume of 0.3 m by 0.38 m by

0.38 m with an accuracy of 0.25 mm to 0.13 mm at a cost of $US289 thousand. Materials such as ABS have been used [13].

BPM and MJM use moving heads to spray fine particles of material onto a lower part. 3D Systems produce an MJM machine with a work volume of 0.20 m by 0.20 m by 0.25 m with an accuracy of 0.08 mm or more, at a cost of $US60 thousand [2]. BPM Technology markets a machine with a work volume of 0.15 m by 0.2 m by 0.25 m with an accuracy of 0.13 mm at a cost of $US34 thousand. [7,2].

Sanders uses an approach much like a 3D dot matrix printer. The print head can deposit either low or high melting-point materials as it scans across the part. The low melting-point material is then melted to leave the high melting-point material. This process has the potential to become a low-cost process.

FDM effectively uses a 3D plotter that moves about an object and extrudes a bead of material. This method is well suited to shells, but the variety of materials can be limited, and the equipment costs have the potential to move into the midrange. Stratasys Inc. markets a modeler with a work volume of 0.3 m by 0.3 m by 0.3 m and an accuracy of 0.13 mm at a cost of $US172 thousand[3].

DDP sprays small jets of adhesive onto a powder base material. This hardens the material selectively, and as layers of powder are added and partially hardened the final part is built. Eventually the loose powder is removed to liberate the completed part. Soligen Inc. sells a machine that has a work volume of 0.4 m by 0.4 m by 0.4 m with an unspecified accuracy, at a cost of $US250 thousand.

MD* is a research-oriented process that uses sprayed metal through masks to form metallic parts [4]. This is of note because it shows an attempt to move towards metal prototypes, and away from the plastics and laminated materials.

At present these processes have a number of difficulties, that are being overcome. All these processes generally produce non-metallic parts from limited ranges of materials. These are not suited to rugged testing. The photopolymer-based systems tend to be very expensive to operate because of high material costs (typically > $US100 per liter). Layer thickness results in a trade-off between surface smoothness and build times. The mechanisms (such as laser optics) tend to limit the work volumes; a cubic foot is considered large of many of these processes. Some of the processes are sensitive to set-up parameters (e.g. humidity and temperature). Some methods require that extra supports be added, requiring extra planning before and after processing begins.

An emerging area of research and development has been techniques for the production of metal parts and tooling directly from various free form fabrication parts. Metal parts can be produced by using techniques such as investment casting [12]. Short run injection moulding tooling can be produced using spray metal coatings on RP parts [16].

11.2.2 Subtractive processes

In subtractive processes we generally begin with a block of material, and using standard machining techniques (i.e. milling, drilling, turning, etc.) we remove material until only the desired part remains. The basic processes are available on most factory floors, and researchers are generating exciting new technologies at all times.

Of greatest note, numerical control machines were first demonstrated in the 1950s but are now commonly used in most industries. It is possible to buy a desktop manufacturing system, including a computer, CAD software, CAM software and a desktop CNC milling machine, to produce small parts for less than US$30 thousand. Giddings & Lewis Inc. sell a mill that can machine volumes of up to 22.4 m by 4.5 m by 4.5 m, to an accuracy of 0.008 mm, and it sells for US$2–8 million. The Light Machines Corporation sells a desktop NC machine with a work volume of 0.3 m by 0.15 m by 0.23 m with an accuracy of 0.013 mm for US$12 thousand [6]. Many of the problems in cutting processes occur because removing material involves significant cutting forces. These cutting forces dictate that the part must be securely fixed, and that the machine tool must be rigid enough to resist deformation. Most subtractive processes have a preferred or fixed work axis. For example, a drill has a single approach axis, whereas a five-axes mill can reach all positions and orientations. To further complicate the situation, the tool may collide with the workpiece (causing gouging, for example) or with the fixture. These factors typically require a high level of human involvement for planning and fixturing.

There are alternative subtractive processes. For example, EDM (electro-discharge machining) removes material using microscopic sparks that melt off small quantities of metal. A wire EDM machine is sold by Sodick Inc. and it has a work volume of 0.75 m by 0.54 m by 0.31 m with an accuracy of 0.005 mm for $US250 thousand [6]. These processes can be very slow, and they require conductive workpieces, but there are no cutting forces.

11.2.3 A case for a compromise

The additive techniques are currently the subject of enthusiastic research and development by the academic and commercial sectors. By comparison the subtractive technologies have matured, and are well understood and respected in traditional industries. Most manufacturers that are considering additive technologies are faced with having to adopt new technological capabilities, design procedures and fabrication processes. It can be hard to justify rapid prototyping, and based on direct cost savings alone many companies would not be able to justify the expense [15]. Some companies are not willing to make this leap of faith for the promises of eventual payback.

We might be able to make the leap to free form fabrication if the initial step were not so large. What most companies would prefer is a technology that employs investments and skill already available in-house, but allows the benefits of the additive technologies. A free form fabrication technique that uses conventional CNC milling machines (note, a subtractive process) would be usable in facilities not willing to commit to additive technologies. There is an obvious conflict if a subtractive process is to be used to perform additive processes. Burns [6, p.7] indicated that there have been no processes to date that combine additive and subtractive processes, except for processes that cut sheets, and then bend them into layers. More recently CAM-LEM was developed [8] using stacked layers of laminates, such as green ceramics, that are cut and then stacked.

11.3 HOW TO PERFORM FREE FORM FABRICATION WITH MILLING MACHINES

Let us consider computer-controlled milling machines that are capable of cutting a wide variety of features accurately. A three-axes mill is capable of reaching all the positions in the workspace of the part, but it is incapable of cutting 'under' the part. Even a five-axes milling machine (that is sufficient to reach the part in any position or orientation) is not suited to cutting 'under' the part where fixtured. In both cases the part would normally need to be partially cut, then refixtured to complete the cutting. If we can make the undercuts first, then put the part in front and then make the overcuts, then this eliminates the problem of tool access to the surface. To do this a layer of disposable material can be deposited, and the undercuts for the final part are made first. This negative then acts as a mold for the workpiece. Once 'cast' into the mold, the undercut surfaces are complete, and the overcuts on the top surface are then made. This technique will result in a part with a full set of over- and undercuts on the external faces.

The physical volume of the tool, chuck and spindle means that it may not reach inside the part to make cuts in hidden, deep or tight locations. This problem may be overcome if the part is built in layers. Moreover when no internal features need to be cut, the layers can be made thicker.

This process is called additive milled pocket (AMP) manufacturing. In the example of Fig. 11.1, the idea of cutting layers is illustrated. Basically, a layer of wax is poured and allowed to solidify. Once stable, a pocket is cut and then filled with a polymer. This is then allowed to harden and another layer of wax is added on top. Once set, the next layer of wax is cut and filled with polymer. This process is repeated until the part is complete, and then the wax is melted to reveal the inner part.

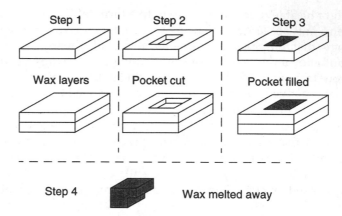

Fig. 11.1   A simple example of the AMP process.

11.4 INVESTIGATING THE METHOD

The main goal of this method was to use a common CNC machine to perform free form fabrication; as a result, testing was planned to explore procedural steps for various features. To date the method has been tested in general. Future work will focus on the implications of material selection and cutting conditions. Testing was divided into two steps, general procedures and normally unmachinable features. First, the general procedures were examined to identify potential problems and the details needed when using the technique. Second, features that cannot be machined easily were considered.

11.4.1 General procedure and results

To determine usability, the process has been tested in a number of machines using the general procedures outlined below (more details are available in [9 and 10]).

1. Examine the part to find the best orientation; try not to have single surfaces that will both under- and overcuts. Divide the part into layers; the layers do not need to not need to be separated by planar sections.
2. Select a container larger than the part to work in, and mount it on the milling machine.
3. Put down a layer of thermoplastic material (wax); it should be thicker than the current layer of the part. Keep in mind that the part is built from bottom to top. A fan can be used to speed the hardening of the wax.
4. Using the CNC mill, cut out the volume of the slice (generally the bottom-facing surfaces of the part). At this point the undercut surfaces should be completed, and some material removed from the layer below. Use a vacuum to remove chips.

5. Fill the cavity with the thermoset polymer (autobody filler), and then allow it to harden. If the mixture is viscous, it may require some working to get it into the corners of the mold. It is advisable to overfill the cavity slightly, as excess material will be removed.
6. Make any overcuts at this point. If any material joins to the layer above, it does not need to be cut even; this will happen when the next layer is added. The vacuum can be used to remove chips during cutting.
7. Repeat steps 3–6 until the part is complete.
8. The part is now encased in wax and this can be melted off. The wax should be kept and reused.
9. The chips that have been collected during cutting can also be melted. The hardened (and now inert) polymer settles out and can be disposed of safely; the wax should be reused.

Generally, early tests used Paraffin wax and autobody filler. Paraffin wax was chosen because it is easy to remove when the part is completed, but can provide enough strength to support and mold the part. The autobody filler was chosen to be dimensionally stable as it sets, and so that it would pour easily, not melt the paraffin while setting, and not melt itself when the paraffin was melted off. Both these materials were purchased in hardware and craft stores.

Early tests were performed using a drill press with an square end mill cutter; this resulted in crude parts. It was noted that when large volumes of bodyfiller were used, the heat from exothermic reaction rose significantly, requiring care to not melt the wax. Other tests were performed using a manual turret mill. The quality of the parts was much higher, but the process was very slow. Dissimilar materials were used to show the capabilities of the process. This was followed with tests made on a large-scale three-axis horizontal CNC mill. The cutting time was very short; the wax setting time was significant, but a fan was used to speed the process. In general it was found that a set of small parts could be produced within a few hours. Recently, two desktop milling machines (these are described later) have been developed to cut lightweight materials, and with special considerations for this process.

A simple test geometry can be seen in Fig. 11.2 that illustrates various techniques for producing a sphere. On the left is the sphere cut with stacks of cylinders. Here the cylinder heights create a staircase effect on the surface. This process is also relatively slow, but it was found to be necessary when the heat generated by large volumes of setting polymer tended to melt the wax. The technique on the right contains two unique features. First the sphere is cut in two steps only, and second by cutting rounded surfaces the sphere is smoother (equivalent to a large number of layers). Moreover, using a ball head cutter gave a much better surface texture. The cutting process was fairly routine, but the wax needed up to

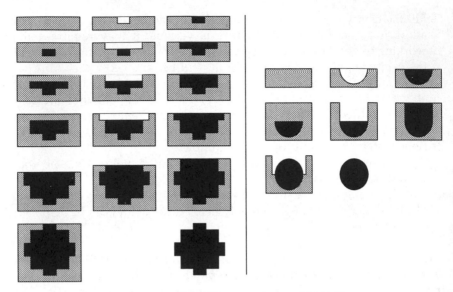

Fig. 11.2 Various techniques for making a sphere with AMP.

½ hour to set for layers a few inches thick. The polymer typically set in about 10–20 minutes; larger volume set faster because of higher temperatures. Other materials have been investigated as replacements for the wax and autobody filler, but these will be described in future publications.

11.4.2 Special machining cases

To determine how this technique extends the normal capabilities of NC machining, thought/practical experiments were conducted. The features considered were identified from a number of sources, including product literature, papers, books and discussion groups.

Small and special features

As feature dimensions become smaller than the cutting tool we may change the strategy of cutting layers. For example, to cut a narrow square hole in a part we may cut only a few of the four sides first, effectively leaving a thin section of wax in the part, and then in a second operation, cut the slot to a small hole. The example of Fig. 11.3 shows another possibility to create a small hole. It is also possible to use special form tools to obtain high-quality features in the final part, such as taps to cut threads. I have also tried pressing shapes for texturing. For example, a coin pressed into the wax will give a raised coin impression on the final part.

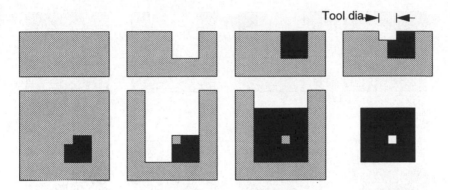

*Note: Using this technique it is possible to make features that are smaller than the cutting tool.

Fig. 11.3 Cutting features smaller than the tool with AMP.

Constructive solids geometry (CSG)

Constructive solids geometry (CSG) parts use addition and subtraction of solids to construct a part. It is possible to create a block of wax larger than the final part. This wax will act as a matrix to support the work. We can then cut cavities into the work and fill it with bodyfiller to add to the part, or add wax to subtract from the final part. Fig. 11.4 below shows the

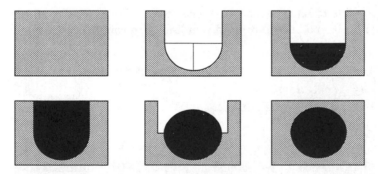

Fig. 11.4 Creation of a CSG primitive with AMP.

creation of a simple primitive shape in the matrix. This procedure would basically involve reduction of the CSG expression to a canonical form, and then selecting one primitive shape at a time to add or subtract. An example of this method can be seen in Fig. 11.5 where a fairly common CSG example is shown here.

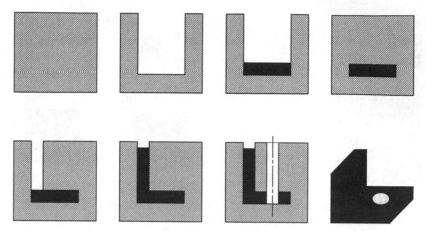

Fig. 11.5   A CSG constructed part using AMP.

Inserts and mixed materials

It is possible for the process to be stopped, and special inserts such as nuts or reinforcement ribs to be added during processing. It is also possible to change the types of material being used in the process; for example, materials of different color could be added. Typical variations that I have tried so far have included glass fibre reinforced plastics (strength), and aluminum-filled plastic (thermal conductivity and metallic appearance).

Assembly and coatings

It is possible to have two parts preassembled. For example, a hole–shaft pair could be made by putting down half of the shaft, applying a parting

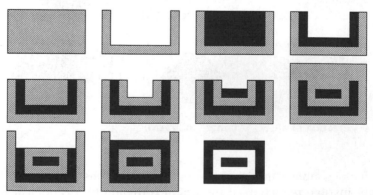

This is a combination of two different parts. The center part could be another material, or an inserted nut.

Fig. 11.6   Mixed materials and/or inserts using AMP.

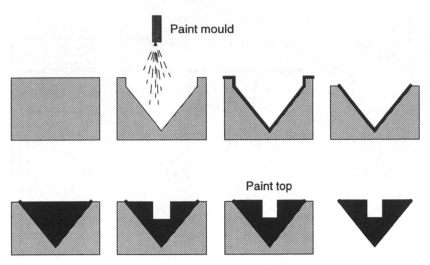

Fig. 11.7 A part painted in the mould using AMP.

agent, then building the shaft in place. Finally the uncollared part of the shaft is coated with parting agent, and the second half of the shaft is constructed. Coatings, such as paint, could be applied to the part. To paint undercut surfaces, the wax negative of the bottom facing surfaces are painted before the bodyfiller is added. Overcut surfaces are painted after they have been cut.

Handy extras

Unlike the methods that are purely additive, or subtractive, mistakes or random failures can be undone. For example, a designer may notice during the build of the part that a feature is round when it should be square. The process could be stopped, and the hole replaced by the correct shape, and cutting could continue. Similar and common problems might be that materials runs out, a flaw is seen in the part, etc. Cutting speeds and tools can be varied to suit various features and surface finishes. The use of lighter-weight materials means that higher speeds and feeds are possible.

11.4.3 A specialized machine

As mentioned before, specialized machines have been designed and built for the AMP process. These were designed with the objective of making the structure lighter because of reduced cutting forces, and they have also endeavored to incorporate the wax and polymer feeders. The basic set-up for one of the fully automated AMP machines is shown in Fig. 11.8 below.

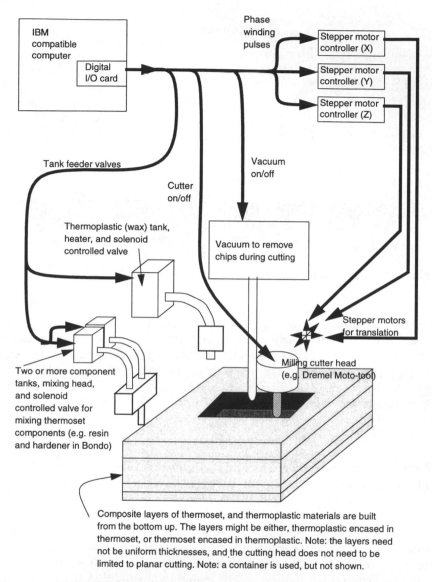

Fig. 11.8 System diagram for a dedicated AMP machine.

This system uses stepper motors for positioning in X and Y by moving the table. To do this the stepper motors drive gear boxes that in turn drive timing belts attached to the table. The Z axis is actuated using a rack and pinion driven mechanism. Because stepper motors are used on all axes, position feedback is not necessary. The vacuum is controlled by a relay that switches 115 VAC. The milling head (a Dremel Moto-tool) has a relay switched 0 V/3 V/6 V supply that allows variable speeds. The cutting head in use is a small hobby handtool that is essentially a hand-held machine spindle. There is a wide variety of toolbits available for this handtool. It is recommended that a cutting tool with a low or negative rake angle be used because of the materials being used, and to reduce the upward forces on the material. The wax/polymer feeders use solenoid-actuated valves and all tanks are under a light pressure to overcome viscosity of the liquids. The wax tank also has a small heater inside to keep the wax in a liquid state.

The sections of the system mentioned above have been completed, but a couple of tasks remain. First, the mixing head for the thermoset polymer is demanding, in that it needs to be able to take two different materials with moderate viscosity (generally 1000–3000 c.p.s.) and mix them thoroughly. The mixed material is then dispensed. The difficulties come in dealing with any residue that will fully harden before the mixing head is used again. This issue is being examined and will be resolved in the near future; another ongoing concern is the control software. At the low level this system requires a driver for the machine using some sort of numerical control code. Control software using the G-Code has been implemented, but at the higher level a solid modeler is required to generate the NC code automatically.

11.4.4 Identified questions

Some preliminary work on optimal cutting conditions for these materials has been done. However, because of the fact that for best utilization of the technique these materials will be cut before they have fully hardened, thus introducing nonlinearities, the search for alternative materials is an ongoing quest. At the time of writing there are dozens of build materials, and a few matrix materials are found to be suitable. The materials also affect the accuracy and repeatability of the process.

11.5 CONCLUSION

In general the AMP process has shown a great deal of potential, and another set of investigations is being planned.

11.5.1 Advantages/disadvantages

The advantages can be summarized as follows:

- thicker layers decrease part production time
- it makes use of an existing base of NC mills
- materials are inexpensive and very easy to obtain

- safety concerns are minimized
- existing machining systems provide strong support
- sizes well above 1 m/1 yd are possible
- parts could be painted during production
- inserts are easy to add
- no assembly is required
- the matrix (e.g. wax) can be collected and reused.

The specific disadvantages of this techniques are that:

- more sophistication is required in the control software
- intermittent mixing of thermoset materials can be difficult
- shrinkage of materials may require dimensional compensation.

11.5.2 Summary

A process has been presented that uses a hybrid of additive and subtractive processes for rapid prototyping of parts. The basis of the operation is wax layers that are poured and allowed to harden. Pockets are then cut into these layers, and a hardening polymer is poured in. Once both are solidified, the tops of both surfaces are milled flat, and a new layer of wax is added. This continues until the entire part is built up, at which time the wax is melted and the encased part is removed. This process can be made in any machine shop using standard milling machines. The total retail costs for consumable supplies could be less than $20 for a simple model, as compared with the initial capital costs of $US35–500 thousand for an additive process, and typically high material costs. Even more importantly, this process also draws on the experience base of existing machinists, and utilizes an existing infrastructure for machine tools.

ACKNOWLEDGMENTS

I must mention my Research Assistants: Rob Tremblay made a major contribution; others include Kevin Moss, Mark Doogan, Andrew Duncan, Jim Vergas and Mark Phoa. This research was made possible through internal support by Dr R. Guerriere and Dr W. White.

REFERENCES

[1] Aronson, R. B. (1996) Spare parts for the over 50 crowd. *Manufacturing Engineering*, Mar., 87–92.
[2] Ashley, S. (1996) Rapid concept modellers. *Mechanical Engineering*, Jan.
[3] Ashley, S. (1996) A new dimension for office printers. *Mechanical Engineering*, Mar., 112–114.

[4] Beck, J. E. (1995) Manufacturing mechatronics using a recursive mask and deposit (MD*) process. Carnegie Mellon University report EDRC 01–20–95.

[5] BPM Technology, Tel. (803) 297–7700.

[6] Burns, M. (1993) *Automated Fabrication; Improving Productivity in Manufacturing*. Prentice Hall.

[7] CAD Systems (1995) *Desk-side Personal Modeler(TM) from BPM*. Kerwill Publications, Mississauga, Ontario, Canada.

[8] Cawley, J. D., Heuer, A., and Newman, W. (1995) Solid freeform fabrication directly into engineered ceramics. *ASM Material Congress*, Cleveland Ohio.

[9] Jack, H. (1995) Fabrication method. US patent application filed October 27th.

[10] Jack, H. (1996) Additive milled pocket manufacturing. *Canadian Society For Manufacturing Engineers Conference Proceedings*, Hamilton, Ontario.

[11] Manufacturing Engineering (1996) Print 3-D models fast. *Manufacturing Engineering*, Jan., 24.

[12] Mueller, T. (1995) Developments in rapid prototyping techniques for die cast parts. *Rapid Prototyping*, Rapid Prototyping Association of the Society of Manufacturing Engineers, 1(4).

[13] Rapid Prototyping Association (1995) Advances in materials. *Rapid Prototyping Newsletter*, Society of Manufacturing Engineers, 1(4), 5–7.

[14] Sims, D. (1995) Molecules at your fingertips. *IEEE Computer Graphics and Applications*, Nov., 14–16.

[15] Sorovetz, T. (1995) Justifying rapid prototyping. *Manufacturing Engineering*, Dec., 25–29.

[16] Wohlers, T. (1995) Rapid prototyping state of the industry 1994–95 worldwide progress report. Rapid Prototyping Association of the Society of Manufacturing Engineers.

[17] Zollikofer, C. P. E., and Ponce deLeon, M. S. (1995) Tools for rapid prototyping in the biosciences. *IEEE Computer Graphics and Applications*, Nov., 48–55.

CHAPTER TWELVE

Growing autofab into the 21st century

Marshall Burns

Automated fabrication started more than 45 years ago with the invention of numerically controlled (NC) machining. In the 1980s several additive techniques were developed. These processes work with photocurable plastic resins, thermoplastic powders, adhesive droplets and other specially devised materials. Together, subtractive and additive autofab now offer modern manufacturers the ability to prototype and produce new designs faster and cheaper than ever before.

Future fabricators will go far beyond today's capabilities to offer higher-resolution and faster object generation. They will expand on the currently available fabrication materials, including metals, advanced composites and possibly even living tissues. Machine prices will move both up and down, promoting use by both heavy industry and casual tinkerers. Users may one day communicate designs and design changes via three-dimensional virtual reality environments, making the machines both easier and fun to use.

How is all of this going to come about? The necessary work is underway in hundreds of industry, university and government laboratories around the world. Research and development in automated fabrication may be viewed in terms of the three interconnected fields shown in Fig. 12.1. The first area, process research, looks at methods for manipulating and inducing structure in solid materials. This typically involves issues in mechanical and chemical engineering, such as the power and scanning speed of a photocuring laser, or the temperature and pressure in an extrusion nozzle.

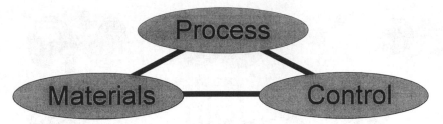

Fig. 12.1   The triad of important issues in automated fabrication research and development.

Materials research studies existing solids and proposes new ones, both from the perspective of their bulk properties (utility after fabrication) and their behavior under manipulative processes (utility as raw material in fabrication). This calls on materials science and chemical engineering for the identification or synthesis of raw materials best suited to a particular process, such as new photopolymer resins, or plastic powders with a specific range of grain sizes. It may also come up with novel materials whose surprising behaviors suggest new fabrication processes.

Control research devises mechanisms and procedures to enact the various fabrication processes automatically. This brings together computer science and mechanical engineering to develop mechanisms, algorithms and protocols.

None of these fields can be studied in isolation from the others. For example, many current autofab processes involve a change in material state from liquid to solid. Optimal control of such processes requires an intimate understanding of the material properties and behavior in the liquid, solid and transition states. Conversely, the final solid properties will benefit from various aspects of control that can be brought to bear on the process.

12.1 PROCESS RESEARCH: LEARNING TO MANIPULATE MATTER

There are three fundamental fabrication processes, as illustrated in Fig. 12.2. In a subtractive process a single piece of material is cleaved in two, whereas additive fabrication causes two separate pieces of material to fuse. A formative process contorts a single piece of material into changing its shape. Each fabrication process, manual or automated, comes down to one or more of these basic processes. For example, die cutting is subtractive, masonry is additive and forging is formative. Spinning pottery on a potter's wheel is a combination of additive and formative fabrication, with an occasional use of subtractive. The most important processes for modern manufacturing, molding processes such as injection molding and investment casting, are formative.

Fig. 12.2 The three fundamental processes are the bases of subtractive, additive and formative fabrication.

Autofab arises when one or more fabrication process is controlled by a computer. Most of today's fabricators are either exclusively subtractive or additive. For example, a CNC mill or lathe uses only subtractive processes. The SLA stereolithography devices from 3D Systems are additive fabricators, as is DTM's Sinterstation. The Helisys LOM, which makes its shapes by bonding and cutting layers of paper, is a hybrid subtractive/additive machine.

The automation of formative processes, which amounts to molding without molds, has attracted only sporadic attention to date. Although many aspects of modern injection molding are highly automated, the process still requires a special tool (mold) to be created for each shape to be molded. This defeats the basic philosophy of automated fabrication, which requires the ability to make arbitrary shapes without special tooling. There has been some rudimentary investigation of changeable-configuration molds, but the first commercial implementation of formative autofab is in CNC press brakes. A press brake is a machine for placing folds and bends in sheet metal. Coupling this with a CNC punch press yields a hybrid subtractive/formative fabricator that turns flat metal sheets into finished desk drawers, refrigerator panels or air-conditioning ducts. Such machines are currently made by Salvagnini and Iowa Precision Industries.

Process research in autofab aims to understand and devise techniques for enacting the three fundamental fabrication processes. The most popular style of cutting breaks off chips with a sharp tool, but other subtractive processes include shearing, abrasion, thermal cutting and chemical dissolution. Joining may use adhesion or cohesion of solid particles, or it may work by bringing together liquid particles and causing them to solidify together. Formative fabrication requires the application of compressive forces on opposing sides of the material at the same time.

Examples of process research include work to develop dual-beam laser curing techniques and options for non-photonic selective curing of plastics. Process research can be motivated by specific goals to be achieved in a fabricator; an example would be working on a method that operates in an open space to defeat limitations of current fabricators to their fairly small build envelopes. One of the most fascinating fields of process research is in nanofabrication: taking the practical scale of fabrication down to the molecular level. Although outside the realm of direct fabrication processes, one could also include in this category robotic techniques intended to take the output objects from a fabricator and assemble them into working mechanisms.

12.2 MATERIALS RESEARCH: GETTING INTIMATE WITH SOLIDS

The properties of a solid material depend on its composition and structure at various levels. The most basic level involves the types of atoms and the ways in which they are arranged (see Fig. 12.3), but at higher levels a polycrystalline material, for example, is affected by the

Fig. 12.3 Three types of solid crystals. In a metal (left) outer electrons float freely about the atomic cores (nuclei and inner electrons represented by black dots). A covalent crystal (center) shares individual electrons between pairs of atoms. In an ionic material (right) outer electrons are shared disproportionately by the different species of atoms present. (Reprinted with permission from *Automated Fabrication – Improving Productivity in Manufacturing* by Marshall Burns, Prentice Hall, 1993.)

sizes, shapes, mutual orientations and the packing density of the crystal grains. Modern chemistry and materials science have done a great deal to identify and explain many types of materials, but surprises continue to arise. Atomic clusters, self-organizing liquid crystals and spherical fullerene molecules are relatively new artificial structures that have forced researchers to rethink our understanding of matter. Of the literally infinite variety of possible composite materials, only a few of the most obviously promising ones, such as fiberglass and reinforced concrete, have been created. Decades, if not centuries, of challenging work remain to be done in exploring the potential for designing new materials for use in autofab.

A very intriguing field at the intersection of materials and process research is that of smart materials. This refers to material systems that respond reversibly to various mechanical, thermal, chemical, or other kinds of influences in their environment. The behavior may involve changes in color, chemical reactivity, physical structure, or other properties. The best example of a smart material is an engineering marvel given to us by nature, called muscle. Our developing ability to create new materials that expand and contract on demand portend a possible future with dynamic or real-time fabrication processes. Such magical processes would create structures whose internal and external shapes are not permanently determined at their time of fabrication, but are subject to a range of movement and reconfiguration.

11.3 CONTROL RESEARCH: INCORPORATING INTELLIGENCE INTO MACHINES

Control is the essence of automated fabrication. It links the mind of the user with the physical processes that create the desired object.

The two most important elements of control are the representation of the desired geometry in computer code, and then the translation of this code into instructions to guide the fabrication process. (see Fig. 12.4). The geometry may arise from a human design, from scanning the shape of an existing object, or from another mathematical source. In the course of fabrication, a good system will monitor its ongoing results and feed them back to the control computer. After the object has been made, the user

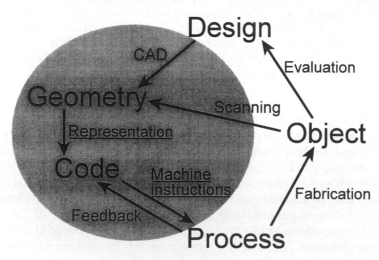

Fig. 12.4 Control aspects of automated fabrication. The key elements, shown here with underlined labels, are the representation of abstract geometry in computer code and the creation of machine instructions to direct the fabrication process.

may evaluate the results and decide whether to make certain changes to the design or to the process parameters.

One very interesting challenge in the arena of control research involves the coordination of multiple process sites. The ability to process material in several places at once offers opportunities for dramatic improvements in speed and efficiency. It is possible that the best way to control such simultaneous processing will use computation devices that are very different from today's serial and parallel computers. Neural networks, fuzzy logic and other new computational techniques may find important applications in controlling advanced fabricators.

At the interface between the user and the fabricator, we can expect to see dramatic improvements in 3D CAD systems. Computer screens, keyboards, and mouses will give way to 3D displays with interactive fingertip point indication and voice instruction. All these technologies are currently under development. Developments in virtual reality technology are making important contributions here. Another element, automatic mechanical property prediction, will allow the user to read the strength of a material in a design in the same way as drawing programs today indicate color on a monochrome screen with various patterns of shading and hatching.

For the more serious fabricator user, there will be matter programming languages. To understand this concept, note that a fabricator is a machine that does with matter what a computer does with information: it takes it in in one form, performs some operations on it, and sends it back out in a different form. A matter language is a system of coded communication that gives a person intimate control over those operations without the person needing to know the details of how the operations are performed. This provides the same advantages that a computer programmer receives from using C or Pascal to achieve efficient, robust processing without needing to be concerned about computer registers and accumulators. Instead of drawing an object graphically in CAD, the matter programmer writes the object in process code. Although casual users will enjoy the convenience of a three-dimensional interactive design environment, professional fabricator engineers and fabricator hackers will appreciate the increased control and versatility available from writing matter programs.

Work underway today in laboratories around the world is making great strides in understanding and improving the processes, materials and control strategies at work in automated fabricators. Although the improvements seen in commercial systems in the last few years are certainly dramatic, these developments have only begun to explore the potential for generating three-dimensional solid objects under computer control.

Index

Bold page numbers refer to figures and italics to tables.